A STOREY ANIMAL HANDBOOK

A GUIDE TO
RAISING
CHICKENS

CARE · FEEDING · FACILITIES

Gail Damerow

STOREY
BOOKS
Schoolhouse Road
Pownal, Vermont 05261

The mission of Storey Communications is to serve our customers by publishing practical information that encourages personal independence in harmony with the environment.

Edited by Pamela Lappies
Cover design by Greg Imhoff and Susan Bernier
Cover photograph by Grant Heilman Photography, Larry Lefever
Technical review by Ralph Ernst
Text design by Cindy McFarland
Text production by Carlson Design Studio and Susan Bernier
Line drawings by Bethany Caskey
Indexed by Gail Damerow

The information in this book is true and complete to the best of our knowledge. All recommendations are made without guarantee on the part of the author or Storey Books. The author and publisher disclaim any liability in connection with the use of this information. For additional information please contact Storey Books, Schoolhouse Road, Pownal, Vermont 05261.

Storey Books are available for special premium and promotional uses and for customized editions. For further information, please call Storey's Custom Publishing Department at 1-800-793-9396.

Printed in the United States by R.R. Donnelley
10 9

Library of Congress Cataloging-in-Publication Data

Damerow, Gail.
    A Guide to Raising Chickens: Care · Feeding · Facilities
           p.   cm. — (A Storey Animal Handbook)

        Includes bibliographical references (p.   ) and index.
        ISBN 0-88266-897-8 (pbk.)
        1. Chickens. I. Title. II. Series.
SF487.D185   1995
636.5—dc20   95-18318
                                     CIP

TABLE OF CONTENTS

Introduction: The Chicken-Raising Mentality

DURING THE SELF-SUFFICIENCY BOOM of the late 1960s and early 1970s, many back-to-the-landers thought they could compete with the poultry industry giants by producing inexpensive meat and eggs at home. Folk wisdom held that chickens could care for themselves by scratching out a romantic existence in the back lot.

When harsh reality set in, however, many of these naive proponents of self-sufficiency found themselves unable to keep up with their flocks' feed demands or to cope with the myriad predators that shared their interest in fresh eggs for breakfast and finger-lickin' chicken for dinner. Disillusioned, they dropped out of home poultry production.

In the 1980s, those who remained loyal to their flocks did so because they enjoyed a sense of self-reliance, they liked the taste of fresh home-grown eggs and meat, and they wanted to control what went into their food. A few continued simply because they got a kick out of keeping chickens. These latter henhouse hangers-on sought not economic rewards but personal satisfaction.

Now the reasons for keeping chickens have once again shifted in emphasis. Natural food production is still a strong motivation, fueled by concern about the safety of our food supply — consumers have grown disgusted with sloppy commercial production practices that let antibiotic residues and resistant strains of bacteria find their way into store-bought eggs and meat. At the same time, poultry preservationists, alarmed about the rapidly vanishing gene pool, are scrambling to preserve endangered breeds. Not to be overlooked are those with apprehensions about the future. When our fragile food distribution system breaks down, they plan to continue enjoying eggs and meat from their own chickens, at least until their flocks are decimated by marauding hordes pouring out of the cities.

Whatever our reasons, we Americans haven't lost what novelist Theodore Dreiser called our "chicken raising mentality" — the great American desire to settle in the country and raise a few chickens. The urge is as strong today as it was in Dreiser's time, nearly a century ago. And rightly so. No other form of livestock combines such a wide variety of colors, shapes, and sizes with low cost and space requirements, offering the opportunity to enjoy an interesting hobby while at the same time putting tasty, healthful food on the table.

CHAPTER 1

CHOOSING THE PERFECT CHICKEN

THE "BEST" BREED for you depends on your reasons for having chickens — for meat or eggs or both, for a touch of beauty in your yard, for show, or strictly for fun. Chickens come in all combinations of laying ability, growth rate, size, shape, and color. You should have no trouble finding the "perfect" chicken — one that's both picture-pretty and ideally suited to your purpose.

Breeds

A *breed* is a group of related chickens having the same general size, shape, and carriage. No one can say for sure exactly how many breeds of chicken there are in the world. The current edition of the *American Standard of Perfection* — commonly referred to as the *Standard* — lists only the 113 breeds recognized by the American Poultry Association (APA), a group that started out as *the* poultry organization but evolved into a promoter of exhibition breeds. (The American Bantam Association publishes its own standard, which doesn't always agree with the APA *Standard*.)

All the birds within a given breed share the same skin color, number of toes, and plumage style. Skin color can be yellow — like the skin of Cornish, New Hampshires, and Wyandottes — or white — like that of Australorps, Orpingtons, and Sussex. They may have four toes, as most breeds do, or five toes, as Dorkings, Faverolles, and Houdans do.

Plumage style is more varied. Roosters are likely to have pointed neck and saddle (lower back) feathers, but if they're Sebrights or Campines, the cocks

1

will be *hen-feathered*, meaning they sport the rounded hackle and saddle feathers of a hen. Naked Necks have no feathers on their necks at all. Other breeds have feathers under their chins that form beards (Faverolles, for example), boots down their legs and feet (Brahmas), puffy top knots (Polish), or long flowing tails (Yokahamas).

Birds of most breeds have smooth, satinlike feathers, a result of the tiny hooks, or barbicels, that hold a feather's webbing together. The feathers of Silkies, though, lack barbicels, making the birds look as if they're covered with fur. The feathers of Frizzles curl at the ends, giving the birds a permed look. Frizzledness is a condition that can pop up in any bred, besides appearing in the breed called "Frizzle."

All breeds are pictured and described in the *Standard*, which divides the breeds into two groups: *bantam* and *large*. Bantam breeds are one-fifth to one-fourth the size of the large breeds. Some bantam breeds are miniature versions of a corresponding large breed; others are distinctive breeds in their own right.

Classes

Each group, large and bantam, is subdivided into a number of classes.

The classifications for large breeds indicate their places of origin: American, Asiatic, English, Mediterranean, Continental, and Other (including Oriental). Each large breed is listed in only one class.

Bantams are classified according to certain characteristics: by whether or not they are game breeds, by their comb style, and by the presence or absence of leg feathering. Among bantams, the same breed may be represented in different classes by distinctive varieties.

Varieties

Most breeds are broken down into *varieties*, categories based usually on color but sometimes on feather placement or comb style.

Plumage color ranges from a rainbow of solid colors to patterns such as speckled, barred, or laced. Wyandottes, for example, come in nine varieties based on color, including solid hues such as buff, black, and white, as well as patterns such as gold or silver lacing.

Varieties defined by feather placement might have, for example, feathers on the legs or under the chin. Frizzles come in two varieties, clean-legged and feather-legged. Polish, Booted bantams, and Silkies may be either bearded or nonbearded.

Breed by Comb Style

Single Comb

Ancona*, Australorp, Blue Andalusian, Campine, Catalan, Cochin, Delaware, Dorking*, Faverolle, Frizzle**, Holland, Java, Jersey Giant, Lakenvelder, Lamona, Langshan, Leghorn*, Minorca*, Modern Game, Naked Neck, New Hampshire, Old English Game, Orpington, Phoenix, Plymouth Rock, Rhode Island Red, Sussex, White Faced Spanish

Rose Comb

Wyandotte

Rose Comb (spiked)

Ancona*, Dominique, Dorking*, Hamburg, Leghorn*, Minorca*, Redcap, Rhode Island Red*

Cushion Comb

Chantecler

Pea Comb

Ameraucana, Araucana, Aseel, Black Sumatra, Brahma, Buckeye, Cornish, Cubalaya, Shamo

Walnut Comb

Yokohama

Strawberry Comb

Malay

Buttercup Comb

Sicilian Buttercup

V Comb

Crevecoeur, Houdan, La Fleche, Polish, Sultan

*Breeds having varieties based on differing comb styles.
**The *condition* of frizzledness may appear in combination with any comb style.

The most common comb style among chickens is the single comb, a series of upright sawtooth zigzags. Varieties defined by comb style might have buttercup, pea, rose, cushion, walnut, strawberry, or duplex (cup or V) combs. Among breeds with varieties defined by comb, Anconas and Rhode Island Reds each have two varieties — single comb and rose comb. Leghorns are an example of a breed that comes in different colors *and* different comb styles; among the possibilities are buff, black, or silver with either single or rose combs.

Strains

A *strain* is a related family of birds bred with emphasis on specific traits. Strains bred by fanciers are usually derived from a single breed, developed for what the owner perceives to be superior appearance. Commercial strains are often hybrids — parented by a hen of one breed and a cock of another — developed for efficient production. Commercial meat strains and brown-egg layers are usually true hybrids. Commercial white-egg layers, on the other hand, aren't hybrids in the strictest sense, since they are crosses between different strains of the same breed and variety — single-comb white Leghorn.

Whether hybrid or *purebred* — parented by a hen and rooster of the same breed — birds within a strain are so much alike that an experienced person can recognize the strain at a glance. An established strain is usually identified by its developer's name — commercial strains by their organization's name, such as Hubbard and Shaver; exhibition strains by an individual's name, such as Halbach. An exhibition strain might not be named at all until its owner dies and the flock is dispersed.

Temperament

Different breeds, and different strains within a breed, vary in temperament, and temperament varies with management style. These extremes in temperament are common in the following breeds:

Docile	Flighty	Aggressive
Cochin	Hamburg	Aseel
Dorking	Lakenvelder	Cornish cocks
Orpington	Leghorn	New Hampshire cocks**
Plymouth Rock	Polish*	Rhode Island Red cocks

*If they have large crests preventing them from seeing well.
**Production strains, closely related to Rhode Island Red.

Breed Selection

Way back in time, all the various breeds, varieties, and strains had a single origin — the wild red jungle fowl of Southeast Asia. Over tens of thousands of years, chicken keepers selectively bred their flocks to favor different combinations of characteristics, some related to economics, some to aesthetics.

Today's breeds, varieties, and strains vary greatly, from common (and therefore inexpensive) to extremely rare (and therefore quite dear). Pure strains of Araucana or New Hampshire, for example, are uncommon to rare, but crossbred production strains are as common as dirt. Not every hen that lays a blue egg is an Araucana, and most hens sold as New Hamps are crossed with Rhode Island Reds or other production breeds. Indeed, among the batch of brown-egg laying "purebred" New Hampshire hens I once bought, some laid blue eggs.

Laying Breeds

The Mediterranean breeds, especially Leghorn, are efficient layers. Good laying hens share four characteristics:

- ◆ They lay large numbers of eggs.
- ◆ They have small bodies.
- ◆ They begin laying at about 5 months of age.
- ◆ They are not inclined to brood.

The best layers average between 250 and 280 eggs per year, although individual birds may exceed 300. Compared to larger hens, these small-bodied birds do not need as much feed to maintain muscle mass. Purebred hens rarely match the laying abilities and efficiency of commercial strains, but they can still be efficient enough for backyard flocks. Since a hen stops laying once she begins to brood, the best layers are nonsetters.

Leghorns, like other Mediterranean breeds, are quite high-strung and therefore not much fun to work with, especially if you've taken up chicken-keeping for relaxation. Another characteristic of the Mediterraneans is that they lay white-shelled eggs.

To satisfy market demand, brown-egg hybrids have been developed that lay as well as Leghorn-based strains. Brown-egg hybrids are derived from breeds in the American classification, which aren't nearly as flighty as the Mediterraneans. One popular brown-egg strain is the Hubbard Golden Comet, a buff-colored bird called "the brown-egg layer that thinks like a Leghorn." You can expect 180 to 240 eggs per year from a commercial strain brown-egg layer.

Large Breeds

Class	Breed	Varieties	Eggs	Meat	Ornamental	Sport	Prevalence
AMERICAN	Buckeye	1	*	*			rare
	Chantecler	2	*	*			rare
	Delaware	1	*	*			rare
	Dominique	1	*	*			rare
	Holland	2	*	*			rare
	Java	2	*	*			rare
	Jersey Giant	2		*	*		rare
	Lamona	1	*	*			rare
	New Hampshire	1	*	*			uncommon
	Plymouth Rock	7	*	*	#		#
	Rhode Island Red	2	*	*			common
	Rhode Island White	1	*	*			rare
	Wyandotte	9		*	*		#
ASIATIC	Brahm	3		*	*		uncom/rare
	Cochin	9		*	*		uncom/rare
	Langshan	2	*	*	*		rare
CONTINENTAL	Campine	2	*		*		rare
	Crevecoeur	1			*		rare
	Faverolle	2	*	*	*		uncom/rare
	Hamburg	6	*		*		uncom/rare
	Houdan	2	*	*	*		rare
	La Fleche	1			*		rare
	Lakenvelder	1	*		*		rare
	Polish	10			*		uncom/rare
ENGLISH	Australorp	1	*	*			common
	Cornish	4		*			uncom/rare
	Dorking	3	*	*			uncom/rare
	Orpington	4	*	*			uncom/rare
	Redcap	1	*		*		rare
	Sussex	3	*	*			uncom/rare
GAME	Old English	13			*	*	
	Modern	9			*		rare

Class	Breed	Varieties	Eggs	Meat	Ornamental	Sport	Prevalence
MEDITERRANEAN	Ancona	2	*				uncom/rare
	Andalusian	1	*		*		uncommon
	Catalana	1	*	*			rare
	Leghorn	12	*		#		#
	Minorca	5			*		uncom/rare
	Sicilian Buttercup	1			*		uncommon
	Spanish	1	*		*		rare
OTHER	Ameraucana	8	*				uncommon
	Araucana	5	*		*		uncom/rare
	Frizzle	2			*		uncommon
	Naked Neck	4			*		rare
	Sultan	1			*		rare
ORIENTAL	Aseel	4			*		rare
	Cubalaya	3			*	*	rare
	Malay	5			*		rare
	Phoenix	2			*		rare
	Shamo	3		*	*	*	uncom/rare
	Sumatra	1			*		rare
	Yokohama	2			*		rare

Not all strains within a breed are equally suitable for the same purpose; nor are all strains equally common or rare. Breeds marked with "#" contain a broader variation than others.

Some breeds originally developed for laying did quite well in their time, but today their laying abilities don't stack up to the production rate of modern strains. Furthermore, many breeds that once had superior laying abilities are now bred strictly for exhibition, where appearance takes precedence over production.

If you have your heart set on a fancy breed but you also want lots of eggs, you have three choices:

◆ Find a strain bred for production as well as appearance (rare).
◆ Develop such a strain yourself, which takes lots of time.
◆ Sacrifice economy for beauty.

After a few years, all layers become "spent," meaning they slow down in production. At that point, they don't have much meat on their bones, since their energy has been concentrated on laying. A good layer fleshes out slowly,

Production Hens

Production hens are either Leghorns (left) or brown-egg hybrids bred for commercial use.

so it never would have made a good meat bird in the first place. If you're interested in chicken barbecue, consider a meat breed.

Meat Breeds

All meat strains share four characteristics:

- ◆ They grow and feather rapidly.
- ◆ They weigh 5 or more pounds by 8 weeks of age.
- ◆ They're broad breasted.
- ◆ They have white feathers for clean picking.

Chickens raised for meat are called broilers, fryers, or roasters. Broilers and fryers are butchered at 3½ to 5 pounds, roasters at 6 to 8 pounds. The more quickly a bird grows to butchering weight, the more tender it is and the cheaper it is to raise.

The most efficient meat strains were developed from a cross between Cornish and an American breed, such as New Hampshire or Plymouth Rock. The one-pound Cornish hens (as a marketing ploy sometimes called "Cornish Game Hens") are nothing more than 4-week-old Rock-Cornish hybrids. A commercial meat bird eats just 2 pounds of feed for each pound of weight it gains. A hybrid layer, by comparison, eats 3 to 5 times as much for the same weight gain.

Raising a backyard meat flock is a short-term project. You buy a batch of chicks, feed them to butchering age, dispatch them into the freezer, and enjoy the fruits of your labor for the rest of the year. Since these birds aren't around long, performance takes precedence over appearance.

The more attractive purebreds may not be as efficient at converting feed to meat as hybrids are, but some are heavy-bodied enough to make good eating. Breeds originally developed for meat include Brahma, Cochin, and Cornish. The Jersey Giant is among the largest breeds but does not make an economical meat bird because it first puts growth into bones, then fleshes out, reaching 6 months of age before yielding a significant amount of meat. Although it wouldn't make sense to raise birds of any pure breed strictly for meat, there's nothing wrong with keeping them for fun or for show and putting surplus birds into the freezer.

Cornish Cross

SHAVER POULTRY BREEDING FARMS LTD.

The Cornish cross is the most efficient chicken for meat production.

Dual-Purpose Breeds

If you want the best of both worlds—eggs and meat—you have two choices: keep a year-round laying flock and raise a batch of meat birds on the side, or compromise by keeping one dual-purpose breed. Dual-purpose chickens don't lay as well as laying hens and don't grow as fast as meat birds, but they lay better than meat birds and grow faster than laying hens. Dual-purpose chickens share four characteristics:

- ◆ They're fairly large-bodied.
- ◆ They're hardy and relatively self-reliant.
- ◆ Most lay large brown-shelled eggs.
- ◆ Many have the instinct to brood.

Dual-purpose breeds are the classic backyard chickens. Their chief advantage over a laying breed is that young excess males and spent layers are full-breasted and otherwise have an appreciable amount of meat on their bones. Their advantage over a meat breed is that the hens lay a reasonable number of eggs for the amount of feed they eat.

Most dual-purpose chickens are in the American and English classifications, but not all breeds in these two classes are dual-purpose. Among the breeds that are, some are slightly more efficient at producing eggs than at growing, some are faster growing and therefore more efficient at producing meat than eggs. These characteristics vary not only from breed to breed, but also from strain to strain within the same breed. As among egg-laying breeds, dual-purpose strains developed for show are generally prettier than they are purposeful.

HOUDAN

BARRED PLYMOUTH ROCK

RHODE ISLAND RED

Dual-purpose breeds are appealing for backyard flocks because they lay better than the meat breeds and grow better than the layer breeds.

Dual-Purpose Chickens

General Purpose	Better for Eggs	Better for Meat
Dominique	Australorp	Black Sex Link*
Houdan	Red Sex Link*	Langshan
Plymouth Rock	Rhode Island Red	New Hampshire
Sussex		
Orpington		
Wyandotte		

*hybrid

To satisfy the market for backyard flocks, a few efficient dual-purpose hybrids have been developed. Most popular of these are the Black Sex Link (a cross between a Rhode Island Red cock and a barred Plymouth Rock hen) and the Red Sex Link (a cross between a Rhode Island Red cock and a white Leghorn hen). The Red Sex Link lays about 250 eggs a year. The Black Sex Link lays slightly fewer but larger eggs, and weighs ¾ pound (0.33 kg) more at maturity.

Ornamental Breeds

In contrast to production birds — chickens kept for meat or eggs — exhibition birds are kept primarily for aesthetic reasons. While the same *breed* might be raised both for show and for production, rarely will you find exhibition and production qualities in the same *strain*. Among the few breeds in which both qualities can be successfully combined are Houdan, Rhode Island Red, and Sussex.

The *Standard* shows and describes the shape, or type, that's ideal for each breed. A chicken that comes close to the ideal for its breed is "true to type" or "typy." Production strains are generally less typy than exhibition strains, since their owners emphasize economics rather than aesthetics. By the same token,

Ornamental chickens are bred for their unique aesthetic qualities, such as the featherless neck of the Naked Neck (top), the curly feathers of the Frizzle (middle), or the startling white face of the Spanish (bottom).

NAKED NECK

FRIZZLE

WHITE FACED BLACK SPANISH

Ornamental Breeds

Breed	Ornamental Characteristic
Ancona	speckled plumage
Andalusian	blue plumage
Araucana	rumplessness, ear tufts
Aseel	upright stance
Brahma	feathered legs
Campine	color pattern
Cochin	basketball shape
Crevecoeur	crest
Cubalaya	long tail, color pattern
Faverolle	bearded, booted
Frizzle	curly feathers
Hamburg	color variations
Houdan	crest
Jersey Giant	huge size
La Fleche	V-shaped comb
Lakenvelder	striking color contrast
Langshan	stately stance, feathered legs
Leghorn	color variations
Malay	upright stance
Minorca	large comb, long wattles
Modern Game	upright stance
Naked Neck	featherless neck
Old English Game	small size, color variations
Phoenix	long flowing tail feathers
Plymouth Rock	color variations
Polish	crest
Redcap	oversized rose comb
Shamo	upright stance
Sicilian Buttercup	cup-shaped comb
Spanish	white face
Sultan	crest
Sumatra	black from head to toe
Wyandotte	color variations
Yokohama	long flowing tail feathers

the typier exhibition strains are less efficient at producing meat and eggs, since their owners emphasize appearance over production.

Even among exhibition strains, not all birds are created equal, since exhibitors emphasize or exaggerate different characteristics. The *Standard* states, for example, that Sebrights should have short, tight backs, yet I've shown Sebrights under a judge who preferred them with long, straight backs — proving the old adage that beauty is in the eye of the beholder.

Endangered Breeds

In 1868 Charles Darwin published an inventory of chicken breeds existing at the time — all 13 of them. Most of the breeds we know today were developed since then. Unfortunately, this incredible genetic diversity has fallen victim to the whims of fowl fads.

Breeds and varieties proliferated in the United States between 1875 and 1925, fueled by interests in both unusual exhibition birds and dual-purpose backyard flocks. The decline began in the 1930s, when new zoning ordinances prohibited chicken flocks in many backyards, and emphasis shifted to commercial production. Among the hardest hit were two breeds with historical significance in North America: Dominique, the oldest American breed, and Chantecler, the oldest Canadian breed.

RARE BREEDS CANADA

The Chantecler has become an endangered breed.

In an ever speedier downward spiral, as birds that had been valued for their appearance declined in popularity, poultry shows became less frequent. As poultry shows became less frequent, interest in chickens declined even further. The depression added its impact — people could no longer afford to keep chickens just because they were pretty.

Interest in exhibiting regained some of its popularity in the affluent 50s, but by then some of the more exotic breeds had all but disappeared. In the spring of 1967 poultry fancier Neil Jones warned *Poultry Press* readers that a concerted effort was needed to preserve these endangered ornamental breeds. Jones's warning led to the birth of the Society for the Preservation of Poultry Antiquities (SPPA), which today attempts to track endangered breeds and varieties with the volunteer help of licensed poultry judges. (The SPPA's address is listed in the appendix.)

Dual-purpose breeds experienced a strong comeback during the self-sufficiency movement of the 60s and 70s, then declined at an alarming rate as people abandoned country life in their scramble for paying jobs. Those of us who stuck with our classic breeds were unaware of the dynamics at work. We butchered our chickens or sold them off by twos and threes, never dreaming they were becoming irreplaceable. While we agonized over the fate of the peregrine falcon and the California condor, we continued taking our domestic chickens for granted.

I count myself among the unwary but guilty. I had a flock of New Hampshires I had worked with for over 10 years. During the 80s I dispersed the birds in preparation for a cross-country move, thinking I could easily replace them when I got resettled. I never again found a strain that equaled my first New Hamps in uniform appearance, rapid growth, and steady laying ability. During my fruitless search, I have heard countless similar stories.

The self-sufficiency shake-out may have sounded the death knell for some of the dual-purpose breeds, but it did save some endangered ornamentals from extinction. The motive of many people who continued keeping chickens changed from one of economics to one of personal satisfaction, and as a result some of the more exotic but less useful breeds were rescued.

In the mid-80s, The American Livestock Breeds Conservancy (ALBC) conducted a survey to identify the most endangered old-time production breeds. In contrast to the relatively informal and subjective SPPA survey, which relied on the observations of a handful of judges, the ALBC survey was somewhat more formal and objective. The ALBC not only identified endangered breeds but located existing flocks and tallied their numbers.

Soon after publishing its first *Poultry Census and Sourcebook* in 1987, the ALBC established the Rare Breed Poultry Conservation Project, which offers

---

Endangered Classics

The ALBC believes that all classic production breeds are in jeopardy but focuses its efforts on twelve varieties having historic, economic, and/or cultural significance:

Ancona

Australorp

Delaware

Dominique

Jersey Giant
  (black and white)

Leghorn (brown)

Minorca (black)

New Hampshire

Plymouth Rock (barred)

Rhode Island Red

Wyandotte (white)

---

assistance to anyone willing to make a long-term commitment to one of its targeted breeds. In an ongoing process, the ALBC continues its review and may, from time to time, add new breeds to its endangered list. (The ALBC's address can be found in the appendix.)

Bantams

Bantams are miniature chickens weighing only one or two pounds. Their history closely follows that of the Industrial Revolution and the movement of families away from farms. Folks who didn't want to give up their chickens turned to miniatures that required little backyard space, didn't eat much, didn't mind being confined, and responded well to human relationships. Interest in keeping bantams boomed in the affluent 1950s, when raising chickens in the backyard was considered good family fun.

Nearly every breed and variety of large chicken has a bantam, or miniature, version one-fifth to one-fourth its size. A few breeds and varieties come only in bantam size. Some people make a distinction between bantams that have a large

POULTRY PRESS

The Japanese bantam is considered a true bantam because it has no large counterpart.

counterpart and those that do not, calling the former "miniatures" and the latter "true" bantams. The *Bantam Standard*, published by the American Bantam Association (ABA), lists more than 350 bantams recognized by the ABA, not all of which are listed in the APA *Standard*.

Banties are popular today as pets, as exhibition birds, and as ornamentals that add a touch of color to the yard. Although their eggs are smaller than the eggs of larger breeds, some banty strains are prolific layers. The chunkier breeds may not rival commercial Rock-Cornish in growth rate, but they look and taste just as good, if not better.

Bantams

Class	*Breed	Miniature/ True Bantam	Varieties**	Prevalence
GAME	Old English	M	19	#
	Modern	M	12	#
SINGLE COMB,	Ancona	M	1	rare
Clean Legged	Andalusian	M	1	rare
(other than game)	Australorp	M	1	uncommon
	Campine	M	2	uncommon
	Catalana	M	1	uncommon
	Delaware	M	1	rare
	Dorking	M	2	rare
	Frizzle	M	1	uncommon
	Holland	M	2	uncommon
	Japanese	T	6	#
	Java	M	2	uncommon
	Jersey Giant	M	2	uncommon
	Lakenvelder	M	2	uncommon
	Lamona	M	1	uncommon
	Leghorn	M	10	#
	Minorca	M	3	#
	Naked Neck	M	4	uncommon
	New Hampshire	M	1	common
	Orpington	M	4	uncommon
	Phoenix	M	2	uncommon
	Plymouth Rock	M	7	#
	Rhode Island Red	M	1	common
	Spanish	M	1	rare
	Sussex	M	3	uncommon

Class	*Breed	Miniature/ True Bantam	Varieties**	Prevalence
ROSE COMB	Ancona	M	1	uncommon
Clean Legged	Antwerp Belgian	T	9	#
	Dominique	M	1	rare
	Dorking	M	1	rare
	Hamburg	M	6	common
	Leghorn	M	6	#
	Minorca	M	2	uncommon
	Red Cap	M	1	uncommon
	Rhode Island Red	M	1	common
	Rhode Island White	M	1	rare
	Rosecomb	T	3	#
	Sebright	T	2	common
	Wyandotte	M	10	#
ALL OTHER COMBS	Ameraucana	M	8	uncommon
Clean Legged	Araucana	M	5	uncommon
	Buckeye	M	1	rare
	Chantecler	M	2	rare
	Cornish	M	5	#
	Crevecoeur	M	1	rare
	Cubalaya	M	3	#
	Houdan	M	2	uncommon/rare
	La Fleche	M	1	rare
	Malay	M	5	rare
	Polish	M	10	#
	Shamo	M	3	rare
	Sicilian Buttercup	M	1	uncommon
	Sumatra	M	1	rare
	Yokohama	M	2	uncommon
FEATHER LEGGED	Booted	T	6	uncommon/rare
	Brahma	M	3	common
	Cochin	M	13	#
	Faverolle	M	2	uncommon
	Frizzle	M	1	uncommon
	Langshan	M	2	rare
	Silkie	T	4	#
	Sultan	M	1	rare

* The same breed may appear in different varieties in more than one class.
** The American Bantam Association may recognize additional varieties.
# Some varieties are less common than others.

Self-Reliant Breeds

Some breeds are inherently more self-reliant than others. Chickens that have been bred in confinement for generations are generally less self-reliant than those that have been allowed to exercise their foraging instinct. In the South, for example, Old English Games are commonly seen wandering along country lanes. As the closest living kin to the ancient wild jungle fowl, they aren't as plump or prolific as the dual-purpose breeds, but they compensate by being almost entirely maintenance free.

Good Foragers

Ancona
Aseel
Campine
Dominique
Hamburg
Leghorn*
Minorca
Old English Game
Orpington
Rhode Island Red
Sussex
Wyandotte

*Non-commercial strains

Two other characteristics that enhance a breed's self-sufficiency are plumage color and brooding instinct. Feather colors other than white blend more easily into the surroundings, offering birds protection from predators. Breeds that have retained their instinct to brood require no human intervention to reproduce, in contrast to their specialized industrial cousins whose brooding instinct has been taken away.

Breeds that are not good foragers or that should not be turned free to forage include those with feathered legs and/or crests. Leg feathering inhibits scratching the ground to turn up food. Crests inhibit vision, making crested breeds easier prey. In freezing weather, breeds with tight combs such as cushion, pea, or rose can cope with the cold better than breeds with large single combs.

Breeds for Feathers

Colorful chicken feathers are prized for tying fishing flies, making jewelry and home decorations, and trimming hats and other clothing. Different crafts require different kinds of feathers. For some crafts, wing and tail feathers are preferred; for fly tying, the most lucrative feather market, only the pointed hackle (neck) and saddle (lower back) feathers are used.

The best feathers for fly tying come from fast-growing hard-feathered breeds in colorful varieties such as barred Plymouth Rock, blue Andalusian, buff Minorca, and silver-penciled Wyandotte. Although entire flocks are selectively bred for feather color and shape, only the hackle and saddle feathers of cockerels have value.

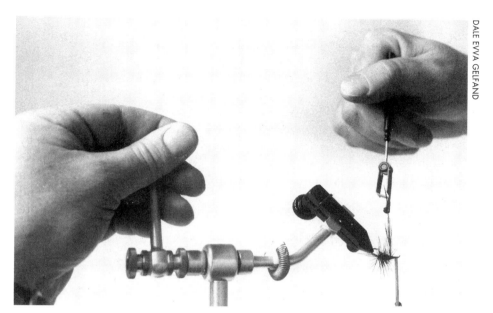

Fly-fishing enthusiasts enjoy tying flies from hackle and saddle feathers of chickens bred specifically for the purpose. A fly-tying vise is used here to secure feathers around a hook.

Fly-tyers use a variety of colorful feathers to recreate the various insects eaten by fish.

When the feathers are prime for harvest, the cockerel is killed and his hackles removed in a *cape* with the skin attached. Saddle feathers are sometimes individually harvested, but are more often removed as a *saddle patch* with the skin intact. Capes and patches are then dried for use or sale.

You can turn a good profit by selling feather products or even marketing the feathers themselves. First you have to find out which feathers are in demand, then how to harvest, process, and package feathers for the market you want to tap. You also need to find out which chickens grow the priciest feathers for that market, and how much the feathers are worth. The most successful feather sellers are associated in some way with the craft for which their feathers are used.

Purebreds vs. Hybrids

Whether you keep purebreds or hybrids depends in good part on whether you wish to incubate your flock's eggs. Purebreds (also called "straightbreds") will breed true, meaning their chicks will be pretty much like the parents. Hybrids won't breed true. Because they result from matings between different breeds (or highly specialized strains within a single breed), the only way to get more chickens exactly like them is to reproduce the cross they came from.

Some Extension agents admonish backyard poultry keepers to purchase a new batch of chicks each year, rather than acquiring replacements by hatching eggs from their own flocks. Their rationale is that repurchasing chicks breaks the disease cycle that otherwise accumulates in a chicken yard. In my experience, just the opposite is true: if you take great care to keep your flock healthy, you're better off hatching from your own birds than constantly bringing in new ones and running the risk of bringing in new problems.

While hybrids are more efficient than purebreds at egg or meat production, they also require expensive high quality feed. By contrast, many purebreds will thrive on forage and table scraps as supplements to commercial rations. They also enjoy a longer productive life.

The decision of whether to keep purebreds or hybrids may depend on whether you intend to show. Except for 4-H shows involving production birds, most shows require entries to conform to their breed descriptions in the *Standard*, meaning the birds must be purebred (although some breeders cheat by crossing their birds with other breeds to improve such things as plumage color and comb type).

Which brings up the question as to whether or not a chicken can be called "purebred" at all. Some people argue that chickens cannot be purebred because they have no registry. Hogwash. Nothing but honor prevents the owner of registered livestock from switching papers or otherwise registering false lineages. The only way *any* animal, registered or not, can truly be considered purebred is through blood typing and DNA testing. Until such testing becomes cheap enough, there's no way to prove a chicken is more or less purebred than any other livestock, registered or not.

Starting Out

Once you know what kind of chickens you want, the next step is to decide whether to purchase newly hatched chicks, partially grown or "started" birds, or fully mature chickens. Each has advantages and disadvantages.

Chicks require additional care at the outset but give you the chance to get acquainted with your birds as they grow. Buying chicks lets you get the most birds for the least cost, and baby birds are least likely to bring a disease into your yard. On the other hand, if you buy exhibition birds, you won't know if you have a potential winner or breeder until the chicks mature.

Chicks come in two options: sexed and unsexed. Unsexed chicks — also called "straight run" or "as-hatched" — are mixed in gender exactly as they hatch, or approximately 50 percent cockerels (males) and 50 percent pullets (females). Some people swear that hatcheries stack the deck by throwing in extra cockerels, since the mix often comes out more like 75 percent cockerels and 25 percent pullets. Sexed chicks are sorted so you can buy exactly as many pullets or cockerels as you want. Within a given breed, sexed pullets cost the most, straight run next, and sexed cockerels the least. Cockerels have the least value because most people have too many of them.

Regulations

Before bringing home your first birds, check your local zoning laws and other ordinances. Regulations may limit or prohibit chicken-keeping activities in your area and may pertain to birds bought, sold, traded, shown, shipped, bred, or hatched. Obtain information from your county zoning board or Extension agent and from your state poultry specialist or veterinarian.

Even when specific laws don't pertain, consider the possibility that grumpy neighbors may file a nuisance suit. If that seems a likelihood, avoid noisy breeds like Ancona, Leghorn, Old English Game, and White Faced Spanish, and avoid free-spirited flying breeds like Hamburg, Leghorn, Old English Game, and most bantams.

If you're establishing a laying flock, your best bet is to buy sexed pullets. For a dual-purpose flock, you might start out with a batch of straight run chicks and raise the surplus cockerels for the freezer. If you're starting a meat project, you can save money on chicks and grow out the birds faster if you get all cockerels.

Started birds, when you can find them, are a good deal if you don't want the bother of brooding chicks. For a laying flock, started—or partially grown—pullets have two advantages: you won't spend much time feeding unproductive birds, since they'll soon begin laying; and, because the birds are just coming into lay, they'll have the longest possible productive life ahead of them. Started birds are also a good option if you plan to show: they're less expensive than proven show birds but less likely to have serious faults than chicks (since birds showing serious faults are culled early on).

Mature birds are the most expensive but offer the fewest surprises, since you see exactly what you're getting. Two unpleasant surprises you can get are disease and excessive age. The older a bird gets, the longer it is exposed to potential diseases, and the more likely it is to carry one. That goes double if the bird has been traveling the show circuit.

Excessive age can be a serious problem if you're buying mature hens for laying or breeding. I once lived near an eggery that sold spent hens to unsuspecting people wishing to start their own backyard flocks. These unwary folks thought they'd get the best hens from a commercial operation specializing in egg production. If you think about it, though, a commercial eggery isn't likely to sell its best layers. The most you can hope for from a place like that is cheap stewing hens. To avoid getting stuck with hens that are past their prime, learn how to tell the difference between a young hen and an old one.

What to Look For

When you buy grown chickens, look for bright eyes; smooth, shiny feathers; smooth, clean shanks (legs); and full, bright combs. To make sure you aren't getting an old, worn-out bird, look for legs that are smooth and clean and a breast bone that is soft and flexible. When you buy chicks, make sure they are bright-eyed and perky. If they come by mail, open the box in front of the mail carrier to verify your refund or replacement claim in case any have died.

A well-kept bird of any age is parasite-free, which you can check by peeking under the wings and around the vent — external parasites can be visibly seen, internal parasites often cause diarrhea that sticks to vent feathers.

If you visit the seller in person, stop to listen for coughing or sneezing in the flock — when a few chickens catch cold, chances are good the whole flock is coming down with it. Old-time poultry keepers whistle whenever they near a flock, causing the birds to pause in their activities so coughs and sneezes are easier to hear.

One way to be sure you're getting healthy birds is to purchase from a flock that's enrolled in the National Poultry Improvement Plan (NPIP). NPIP flocks are certified to be free of two once-widespread diseases, pullorum and typhoid. Some are also free of a still-common disease, mycoplasmosis. Because many breeders don't want to get caught up in the government bureaucracy, you may not find an NPIP member who has the kind of birds you want. (Check the appendix for information on obtaining the latest NPIP directories.)

Flock Size

One of the most common mistakes made by novice chicken owners is getting too many birds too fast. An extreme example is a young couple I knew who had the noble idea of setting up a chicken zoo where they would display every known breed. Before their facilities were ready, they went around buying chickens and crowding them together in a holding pen. The exciting venture soured when, within a few months, most of the chickens got sick and died.

Decide how many birds you want or need, build your facilities accordingly (or a little larger, in case you catch "chicken fever" and wish to expand), acquire the number of birds you planned on, and keep your flock that size. When you buy chicks, get at least 25 percent more than you want to end up with to allow for natural deaths and for culling (elimination of undesirable birds).

If you're starting a laying flock, decide how many eggs you want and size your flock accordingly. As a rough average, you can expect two eggs a day for each three hens in your flock. Since hens don't lay at a steady rate year-round, you may sometimes have more eggs than you can use, and at other times, too few. (See Chapter 7 to learn how to store surplus eggs for lean times.)

If you plan to breed show birds, a mature trio or quartet will give you a nice start. A trio consists of one cock and two hens; a quartet is one cock and three hens.

Unless you're raising cockerels for meat or feathers, most of the chickens in your flock should be hens. If you have too many roosters, they'll fight (see "Female/Male Mating Ratios" on page 184). If you don't need fertilized eggs for hatching, or if the local zoning ordinance doesn't allow roosters, you don't need cocks at all — but you'll miss out on their amusing antics.

Sources

The best place to buy birds depends on what you want. If it's a commercial hybrid strain, your only choice is a hatchery. Unfortunately, some hatcheries churn out large numbers of low quality chicks. The same can be true of chick brokers — feed stores and mail-order houses that market chickens from outside sources, often with little knowledge of or concern for the birds' condition or bloodlines.

Some hatcheries specialize in exhibition breeds, but rarely do they sell prize-winning strains. If you want quality purebreds, look for a serious breeder who keeps records on breeding, production, and growth. If possible, make a personal visit so you can ask questions, examine records, and see the condition under which the birds live.

If you can't find someone locally who has the birds you want, seek out a reputable seller willing to ship. The newspaper *Poultry Press* (listed in the appendix) offers monthly commentary on who's winning at shows and who has birds for sale. Ask exhibitors and judges at poultry shows for tips on who to deal with and, just as important, who to stay away from.

Chicks Shipped by Mail

When you order chicks by mail, open the box in front of the mail carrier to verify any claim you may have for losses. Introduce the chicks to household pets to let them know the chicks are yours and shouldn't be touched. Provide the chicks with heat and water as soon as possible after their arrival.

If you're looking for a classic production strain, the American Livestock Breeds Conservancy (listed in the appendix) can help you find a producer. Seek one who specializes in the specific chickens you want, has worked with the same flock for a long time, and has taken the trouble to trace the flock's history to determine that it came from an original strain.

Time of Year

A good time to visit poultry breeders and examine their flocks is in late November or early December, when young birds are nearly grown and old birds have finished molting. A good time to brood chicks is in March or April, when the weather is turning warm but is still cool enough to discourage diseases. Large breeds started in December and bantams started in March will feather out in time for fall exhibitions. Spring pullets will start laying in the fall and will continue laying throughout the winter.

CHAPTER 2

HOUSING

CHICKEN COOP DESIGNS are as varied as people who keep chickens. The best design for you depends on how many chickens you keep, your purpose in keeping them, their breed, your geographic location, and how much money you want to spend. A handy way to get ideas is to find successful chicken keepers in your area — or correspond with those who live in a similar climate — and pick their brains as to what works and what doesn't.

Coop Design

Some people provide their flocks with perfectly adequate housing by converting unused toolsheds, dog houses, or camper shells. Others go all out, such as the fellow I knew in California who built a two-story structure, complete with a cupola, for his fancy bantams.

No matter how it's designed, a successful coop has these twelve characteristics:
+ is easy to clean
+ has good drainage
+ protects the flock from wind and sun
+ keeps out rodents, wild birds, and predatory animals
+ provides adequate space for the flock size
+ is well ventilated
+ is free of drafts
+ maintains a uniform temperature
+ has a place where birds can roost
+ has nests that entice hens to lay indoors
+ offers plenty of light — natural and artificial
+ includes sanitary feed and water stations

Simple, open housing is easier to clean than a coop with numerous nooks and crannies. If your coop is tall enough for you to stand in, you'll be inclined to clean it as often as necessary. If you prefer a low coop (for economical reasons or to retain your flock's body heat in a cold climate), design the coop like a chest freezer, with a hinged roof you can open for cleaning.

The coop should have both a chicken-sized door and a people-sized door. The chicken door can be a 10-inch-wide by 13-inch-high (25 x 32.5 cm) flap cut into a side wall, opening downward to form a ramp for birds to use when they enter and exit. To keep predators out, the door should have a secure latch you can fasten shut in the evening, after your chickens have gone to roost.

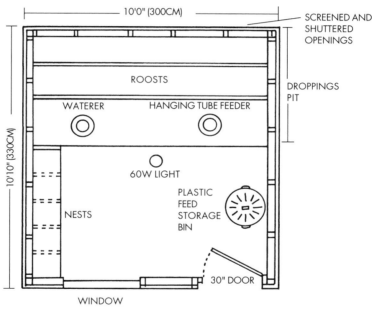

This basic coop plan features roosts over a droppings pit for good sanitation, a window for light, and screened and shuttered openings on the north side to control ventilation. To expand the interior floor space, build the nests on the outside of the coop.

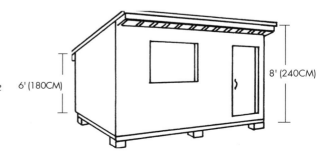

If your soil is neither sandy nor gravelly, locate your coop at the top of a slight hill or on a slope, where puddles won't collect when it rains. A south-facing slope, open to full sunlight, dries fastest after a rain. Capture the sun's light and warmth inside the coop with windows on the south side.

Space

The more room your chickens have, the healthier and more content they'll be. Except in extremely cold climates, home flocks are rarely housed entirely indoors but have room to roam outside whenever they please. Yet even in the best of climates, chickens may sometimes prefer to remain indoors due to rain, extreme cold, or extreme heat.

Minimum space requirements, including those shown in the accompanying chart, indicate the least amount of indoor space birds need when they can't or won't go outside for an extended period of time. Birds that never have access to an outside run will do better if you give them more space than the absolute minimum. On the other hand, birds that spend most of their time outdoors, coming in only at night to roost, will do nicely with less space.

Minimum Space Requirements

Birds	Age	Open House sq ft/Bird	Birds/sq m
Heavy	1 day – 1 week	-	-
	1 – 8 weeks	1.0	10
	9 – 15* weeks	2.0	5
	15 – 20 weeks	3.0	4
	21 weeks and up	4.0	3
Light	1 day – 1 week	-	-
	1 – 11 weeks	1.0	10
	12 – 20 weeks	2.0	5
	21 weeks and up	3.0	3
Bantam	1 day – 1 week	-	-
	1 – 11 weeks	0.6	15
	12 – 20 weeks	1.5	7
	21 weeks and up	2.0	5

* or age of slaughter

From: *The Chicken Health Handbook*, by Gail Damerow (Storey Publishing, 1994)

To encourage chickens to spend most of their daytime hours outdoors, even in poor weather, give them a covered area adjoining the coop where they can loll out of rain, wind, and sun. Encouraging your chickens to stay out in the fresh air has two advantages: they will be healthier, and their coop will stay cleaner.

Ventilation

The more time chickens spend indoors, the more important ventilation becomes. Ventilation serves six important functions:

- supplies oxygen-laden fresh air
- removes heat released during breathing
- removes moisture from the air (released during breathing or evaporated from droppings)
- removes harmful gasses (carbon dioxide released during breathing or ammonia evaporated from droppings)
- removes dust particles suspended in the air
- dilutes disease-causing organisms in the air

Confined Housing		Caged	
sq ft/Bird	Birds/sq m	sq in/Bird	sq cm/Bird
0.5	20	(Do not house heavy breeds on wire.)	
2.5	4		
5.0	2		
7.5	1.5		
10.0	1		
0.5	20	25	160
2.5	4	45	290
5.0	2	60	390
7.5	1.5	75	480
0.3	30	20	130
1.5	7	40	260
3.5	3	55	360
5.0	2	70	450

Compared to other animals, chickens have a high respiration rate, causing them to use up available oxygen quickly while at the same time releasing large amounts of carbon dioxide, heat, and moisture. As a result, chickens are highly susceptible to respiratory problems. Stale air inside the hen house makes a bad situation worse — airborne disease-carrying microorganisms become concentrated more quickly in stale air than in fresh air.

Ventilation holes near the ceiling along the south and north walls give warm, moist air a place to escape. Screens over the holes will keep out wild birds, which may carry parasites or disease. Drop-down covers, hinged at the bottom and latched at the top, let you close off ventilation holes as needed.

During cold weather, not only do you have to provide good ventilation but you also have to worry about drafts. Close the ventilation holes on the north side, keeping the holes on the south side open except when the weather turns bitter cold.

Cross ventilation is needed in warm weather to keep birds cool and to remove moisture. The warmer air becomes, the more moisture it can hold. In the summer, leave all the ventilation holes open and open windows on the north and south walls. Windows should be covered with ¾-inch screen to keep out wild birds and should slide or tilt so they can be easily opened. Provide at least 1 square foot of window for each 10 square feet of floor space (in metric: 1 sq m of window per 10 sq m of floor space).

Where temperatures soar during summer, you may need a fan to further improve ventilation. Hen-house fans come in two styles: ceiling mounted and wall mounted. The former needs be no more than an inexpensive variable-speed Casablanca (paddle) fan to keep the air moving. A paddle fan benefits birds only if ventilation holes are open to keep hot air from getting trapped against the ceiling.

A wall-mounted fan sucks stale air out, causing fresh air to be drawn in. The fan, rated in cubic feet per minute, or cfm, should move 5 cubic feet (0.15 cu m) of air per minute *per bird*. If your flock is housed on litter, place the fan outlet near the floor, where it will readily suck out dust as well as stale air. Since some dust will stick to the fan itself, a wall-mounted fan needs frequent cleaning with a vacuum and/or pressure air hose.

Temperature Control

A chicken's body operates most efficiently at temperatures between 70°F and 75°F (21–24°C). For each degree increase, broilers eat 1 percent less, causing a drop in average weight gain. Egg production may rise slightly, but eggs become smaller and have thinner shells. When the temperature exceeds 95°F (35°C), birds may die.

To keep the coop from getting too hot, treat the roof and walls with insulation, such as 1½-inch styrofoam sheets, particularly on the south and west sides. Cover the insulation with plywood or other material your chickens can't pick to pieces. To reflect heat, use aluminum roofing or light-colored composite roofing and paint the outside of the coop white or some other light color. Try to maintain grass around the coop, keeping it mowed to a height of no more than 4 inches (10 cm). Plant trees or install awnings to shade the building. An awning or other cover can serve the additional purpose of providing a shady, breezy place for birds to rest.

> ### Ventilation Quick Check
>
> Use your nose and eyes to check for proper ventilation. If you smell ammonia fumes and see thick cobwebs, your coop is not adequately ventilated.

To enhance heat retention in winter, build the north side of your coop into a hill or stack bales of straw against the north wall. Where cold weather is neither intense nor prolonged, double-walled construction that provides dead-air spaces may be adequate to retain the heat generated by your flock. In colder weather, you'll need insulation and, to keep moisture from collecting and dripping, a continuous vapor barrier along the walls. Windows on the south wall supply solar heat on sunny days but should be shaded in hot weather.

Flooring

Hen-house flooring can be one of four basic kinds:

Dirt is cheapest and easiest to "install," but consider it only if you have sandy soil to ensure adequate drainage. In warm weather dirt helps keep birds cool, but in cold weather it draws heat away. A coop with a dirt floor is not easy to clean and cannot be made rodent proof.

Wood offers an economical way to protect birds from rodents, but only if the floor is at least 1 foot (30 cm) off the ground to discourage mice and rats from taking up residence in the space beneath it. Wood floors are hard to clean, especially since the cracks between boards invariably get packed with filth.

Droppings boards of sturdy welded wire or closely spaced wooden battens allow droppings to fall through where chickens can't pick in them. Not only will the chickens remain healthier, but droppings will be easier to remove because they won't get trampled and packed down. Start with a wooden framework and to it fasten either welded wire or 1 x 2 inch lumber, placed on edge for rigidity, with 1-inch gaps between boards (in metric: 2.5 cm x 5 cm, with 5 cm gaps). Build manageable sections you can easily remove, so you can take them outdoors and clean them with a high pressure air or water spray and dry them in direct sunlight. Like wood flooring, droppings boards must be high enough off the ground to discourage rodents.

Concrete, well finished, is the most expensive option but also the most impervious to rodents and the easiest to clean. As a low-cost alternative, mix one part cement with three parts rock-free (or sifted) dirt, and spread 4 to 6 inches (10–15 cm) over plain dirt. Level the mixed soil and use a dirt tamper to pound it smooth. Mist the floor lightly with water and let it set for several days. You'll end up with a firm floor that's easy to maintain.

Bedding

Bedding, scattered over the floor or under droppings boards, offers numerous advantages: it absorbs moisture and droppings, it cushions the birds' feet, and it controls temperature by insulating the birds from the ground.

Good bedding, or litter, has these eleven properties:

- ◆ is inexpensive
- ◆ is durable
- ◆ is lightweight
- ◆ is absorbent
- ◆ dries quickly
- ◆ is easy to handle
- ◆ doesn't pack readily
- ◆ has medium-sized particles
- ◆ is low in thermal conductivity
- ◆ is free of mustiness and mold
- ◆ has not been treated with toxic chemicals
- ◆ makes good compost and fertilizer

Of all the different kinds of litter I've tried over the years, wood shavings (especially pine) remain my favorite because they're inexpensive and easy to manage. Straw, unless it's chopped, mats too easily and, when it combines with manure, creates an impenetrable mass. Of the kinds of straw, wheat is the best, followed by rye, oat, and buckwheat, in that order. Any of these, chopped and mixed with shredded corn cobs and stalks, makes nice loose, fluffy bedding.

Rice hulls and peanut hulls are cheap in some areas, but neither is absorbent enough to make good litter. Dried leaves are sometimes plentiful, but they pack too readily to make good bedding. If you have access to lots of newsprint and a shredder, you've got the makings for inexpensive bedding that's at least as good as rice or peanut hulls, but tends to mat and to retain moisture; in some areas, shredded paper is sold by the bale.

If you don't use droppings boards, start young birds on bedding a minimum of 4 inches (10 cm) deep and work up to 8 inches (20 cm) by the time the birds are mature. Deep litter insulates chickens in the winter and lets them burrow in to keep cool in the summer.

When litter around the doorway, under the roosts, or around feeders becomes packed, break it up with a hoe or rake. Around waterers or doorways, remove wet patches of litter and add fresh dry litter (if necessary, fix the leak).

If you use droppings boards, after each cleaning spread at least 2 inches (5 cm) of litter beneath the boards to absorb moisture from droppings. An easy-to-manage combination is to place droppings boards beneath perches where the majority of droppings accumulate, and have open litter everywhere else. Your chickens won't be able to get to the manure piles beneath the droppings boards, but can dust and scratch in the open-litter area, stirring up the bedding and keeping it light and loose. To encourage scratching, scatter a handful of grain over the litter each day and let the flock scramble for it. (For more on litter management, see page 244.)

Roosts

Wild chickens roost in trees. Many of our domestic breeds are too heavy to fly up into a tree, but they like to perch off the ground nevertheless. You can make a perch from an old ladder or anything else strong enough to hold chickens and rough enough for them to grip without being so splintery as to injure their feet. If you use new lumber, round off the corners so your chickens can wrap their toes around it. Plastic pipe and metal pipe do not make good roosts; they're too smooth for chickens to grasp firmly. Besides, given a choice, chickens prefer to roost on something flat, like a 2 x 4 (5 x 10 cm).

The perch for regular-sized chickens should be about 2 inches (5 cm) across, or no less than 1 inch (2.5 cm) for bantams. Allow 8 inches (20 cm) of perching space for each chicken, 10 inches (25 cm) if you raise one of the larger breeds. If one perch doesn't offer enough roosting space, install additional roosts. Place them 2 feet (60 cm) above the floor and at least 18 inches (45 cm) from the nearest parallel wall, and space them 18 inches (45 cm) apart. If floor space is limited, step-stair roosts 12 inches (30 cm) apart vertically and horizontally, so chickens can easily hop from lower to higher rungs. Either way, make perches removable for easy cleanup and place droppings boards beneath them.

Roosts over Droppings Pit

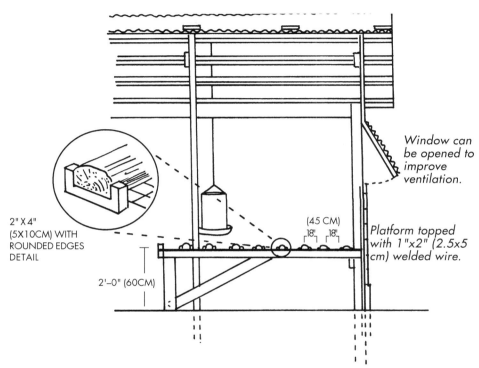

Window can
be opened to
improve
ventilation.

2" X 4"
(5X10CM) WITH
ROUNDED EDGES
DETAIL

(45 CM)
18'   18'

Platform topped
with 1"x2" (2.5x5
cm) welded wire.

2'–0" (60CM)

This roost is made from 2" x 4" (5 x 10 cm) with rounded edges, mounted for easy cleaning and spaced 18" (45 cm) apart over a raised platform surrounded by wire mesh to keep chickens from picking in their droppings.

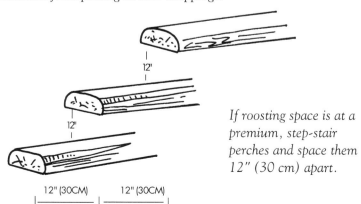

12'

12'

12" (30CM)   12" (30CM)

If roosting space is at a premium, step-stair perches and space them 12" (30 cm) apart.

Nests

Hens, by nature, like to lay their eggs in dark out-of-the-way places. Nest boxes encourage hens to lay eggs where you can find them and where the eggs will stay clean and unbroken. Furnish one nest for every four hens in your flock. A good size for Leghorn-type layers is 12-inches wide by 14-inches high by 12-inches deep (30 x 35 x 30 cm). For heavier breeds, make nests 14-inches wide by 14-inches high by 12-inches deep (35 x 35 x 30 cm); for bantams, 10-inches wide by 12-inches high by 10-inches deep (25 x 30 x 25 cm).

A perch just below the entrance gives hens a place to land before entering, helping keep the nests clean. A 4-inch (10 cm) sill along the bottom edge of each nest prevents eggs from rolling out and holds in nesting material. Pad each nest with soft clean litter and change it often.

Place nests on the ground until your pullets get accustomed to using them, then firmly attach the nests 18 to 20 inches (45–50 cm) off the ground. Raising nests discourages chickens from scratching in them, possibly dirtying or breaking eggs. Further discourage non-laying activity by placing nests on the darkest wall of your coop. Construct a 45-degree sloped roof above nests to keep birds from roosting on top. Better yet, build nests to jut outside the coop and provide access from the back — chickens won't be able to roost over nests, they'll have more floor space, and you'll be able to collect eggs without disturbing your flock.

An alternative plan for creating darkened nests that are easy to clean is to place a long bottomless nest box on a shelf. Partition the inside of the box into a series of nesting cubicles, their entrances facing the wall. Allow an 8-inch (20 cm) gap between the wall and the entrances so hens can walk along the shelf at the back. Build a sloped roof above the shelf to prevent roosting. Add a drop panel at the front of the box for egg collection. To clean the nests, make sure no eggs or hens are inside, then pull the box off the shelf and the nesting material will fall out.

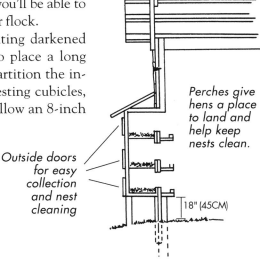

Perches give hens a place to land and help keep nests clean.

Outside doors for easy collection and nest cleaning

18" (45CM)

Exterior nests increase floor space and are easy to maintain from outside the coop.

Nest Boxes on Shelves

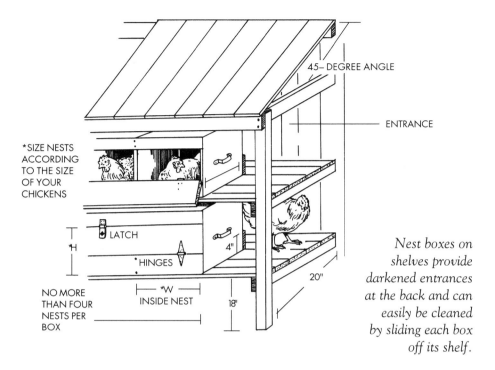

*SIZE NESTS ACCORDING TO THE SIZE OF YOUR CHICKENS

45– DEGREE ANGLE

ENTRANCE

LATCH

H

*HINGES

4"

20"

NO MORE THAN FOUR NESTS PER BOX

*W INSIDE NEST

18'

Nest boxes on shelves provide darkened entrances at the back and can easily be cleaned by sliding each box off its shelf.

Cages

Chickens may be housed in cages for any number of reasons. Commercial laying hens are caged so their diet can be controlled and so they will be protected from diseases, predators, and the weather; layer cages have sloping wire floors so eggs will roll to the outside, where they remain clean and easy to collect. Cocks may be caged to keep them from fighting with one another and/or harassing hens. Exhibition birds are caged for some of the same reasons, as well as to control breeding and condition birds for the showroom.

Caging birds can often be less expensive than building a coop since the cages can be kept in an existing structure. I used to raise exhibition bantams in cages in our garage, where I could control their diet for peak health and production, and where I could be sure their valuable eggs would not be hidden, soiled, or eaten by predators. The birds themselves were protected from predators, provided I closed the garage door at night so vagrant dogs couldn't get under the cages and bite off the birds' feet. To find the cages I wanted, I scoured the classified section until I located a rabbitry going out of business. I picked up all the cages I needed for a song.

Even ready-built new cages may be cheaper than making your own, unless you have an inexpensive source for wire. You'll need 1-inch by 1-inch (2.5 x 2.5 cm) or 1-inch by 2-inch (2.5 x 5.0 cm) galvanized 12- or 14-gauge welded wire. You'll also need cage clips, or "ferrules," ferrule-closing pliers, and wire side cutters. If you buy used cages, you may have to redesign them, for which you'll still need the clips, pliers, and cutters.

First use the chart on page 38 to determine what size cage you need. From welded wire cut four sides, a top, and a bottom, and clip them together along the edges. Around the top and bottom edges, add lengths of 10-gauge wire as reinforcement to keep the cage from sagging.

To make a door, cut a 14-inch (35cm) square opening in the center of the front wall, 2 inches (5cm) from the bottom. File the cut ends smooth. From a separate piece of wire, cut a door 14-inches (35cm) high and 15-inches (37.5cm) wide. Using loose ferrules as hinges, attach the door at the bottom, side, or top — the hinge position is strictly a matter of preference, though a door hinged at the bottom so it drops down when it's opened will leave your hands free. Latches can be fashioned from all sorts of things, but nothing beats the standard cage-door latch in the illustration.

Cage and Tools

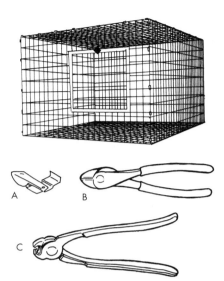

To build or redesign a cage, you will need (a) a latch, (b) wire side cutters, and (c) ferrule-closing pliers, as well as the wire and clips.

Partially cover the cage bottom with a board or piece of heavy cardboard so the birds can rest with their feet off the wire. A resting place is especially important for heavy breeds, since their weight tends to press their feet into the wire. At least once a week, scrape droppings off the board or replace it with a clean one. Lighter breeds appreciate a roost 6 inches (15 cm) off the cage floor, in which case you may have to make the cages 6 inches (15 cm) higher than otherwise so the birds won't rub their combs or topknots against the ceiling.

Cages can be set on sturdy wooden frames or hung from the ceiling in such a way that they won't swing when the birds move around inside. At one time I made cage legs out of concrete blocks, set on end. Today I clip cages to the wall with sturdy picture hangers, which allows for easy cleaning beneath them.

Removing such a cage to transport a bird, or to put it on the lawn to graze through the wire floor, is equally easy.

Cage Dimensions

Number of Birds	Width	Depth	Height*
1	30 in 75 cm	24 in 60 cm	24 in 60 cm
2	27 in 68 cm	32 in 80 cm	24 in 60 cm
4	46 in 115 cm	32 in 80 cm	24 in 60 cm

*Add 6" if you plan to install roosts.

Yard

A yard offers chickens a safe place to get sunshine and fresh air. Ideally, it should have trees or shrubs for shade, along with some grass or other ground cover. Since chickens invariably decimate the ground cover immediately around their housing, a large yard is better than a small one, but a small one is better than none at all.

Some chicken keepers contend that confining a flock to a coop with properly managed litter and good ventilation is more healthful than letting them into a yard of packed dirt coated with chicken manure. That's certainly true if the hardpan turns to slush in rainy weather. Sad but true, two sure signs of an unsanitary yard are bare spots and mud holes.

Where space for a run is limited, one way to avoid the barren-yard problem is to level the area and cover it with several inches of sand. Go over the sand every day with a grass rake to smooth out dusting holes and remove droppings and other debris. Where available yard space is truly minuscule, you might build your chickens a sunporch with a slat or wire floor, and periodically clean away the droppings that accumulate beneath the porch.

At the other end of the scale, if you have plenty of room, keep your flock healthy and take advantage of the cost savings in feed by letting your chickens graze.

Range Rotation

It amazes me that folks who wouldn't dream of planting cabbages or potatoes in the same plot two years in a row never think twice about keeping chickens in the same spot year after year. Even if the coop itself is in constant use — cleaned and disinfected regularly — pathogens and parasites become concentrated in the soil of a constantly used yard.

One way to rotate range is to have two yards with access from the same doorway, through a gate that can be repositioned so when one yard is opened, the other is blocked off. The disadvantage to this system is that the constant comings and goings of chickens through the single entryway soon kills the grass around the door. When it rains, you have an unsightly and unsanitary situation.

To avoid this mess, you might provide different entries into different yards. By having chicken-sized doors on different sides of the coop, you can periodically close one door and open another. As soon as you switch the chickens to a new yard, rake over their previous entryway, toss on some fresh seed, and let the grass grow while the chickens are away. Rest, sunshine, and plant growth will conspire to sanitize the yard.

If chickens must constantly use the same entry, a covered dooryard (of the sort used to provide a shady resting place) helps prevent muddy conditions. Alternatively, put a concrete apron in front of the door and periodically scrape off the muck with a flat shovel. As a third alternative, avoid permanent housing altogether and turn your flock out to pasture.

Range Shelter

Pasturing chickens on range saves you money by letting your chickens forage for much of their sustenance, and it keeps them healthy by preventing a build-up of parasites and pathogens. But ranging requires a fair amount of

Simple Range Shelter

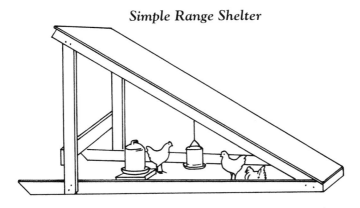

A range shelter to protect the birds from sun and wind can be as simple as a roof on posts. This design can be made from a sheet of plywood covered with rolled roofing or, for lighter weight, from metal or fiberglass corrugated roofing.

ground. It's also labor intensive, since range housing must be moved frequently to new forage areas. Because range housing is moved often, it needs to be light and portable, thus offering little protection against cold weather.

A rudimentary range shelter protects birds from sun and wind. It might be as simple as a roof on posts that can be lifted and moved by two people. For a one-person operation, wheels at one end let you lift the other end and push or pull the shelter to its new location. To protect birds from the elements, enclose those sides of the shelter most subject to prevailing wind and rain.

For our range shelter, we built a plywood corral topped with a lightweight surplus camper shell. Our camper-top shelter is an improvement over open shelters, since it can be closed up at night for predator protection and insulated against cold weather. Although we bolted it together so it could be taken apart for storage or moving, we've found it more convenient to move the whole thing in one piece. For the purpose, we added two sturdy hooks on each side. By using straps or chains to connect opposite pairs of hooks, we can lift and move the shelter with our front-end loader. If you don't have a loader handy, a pair of wheels at one end or skids at the bottom of the shelter would help you move it without too much trouble.

Chickens can be decidedly stupid about finding their home after it's been moved. You can help them along by watching for stragglers that insist on

ALLAN DAMEROW

A plywood corral topped with a lightweight camper shell is easy to move. Also, it can be closed at night for predator protection and insulated against cold weather.

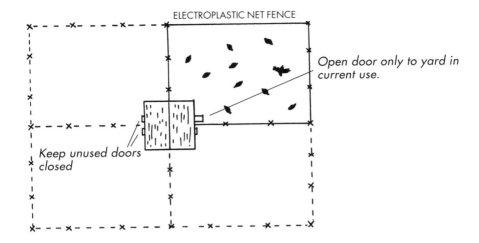

ELECTROPLASTIC NET FENCE

Open door only to yard in current use.

Keep unused doors closed

To rotate range without moving the housing, put chicken-sized doors on different sides of the coop.

bedding down in the old place, and by never moving the house far outside the previous range — chickens are conservative by nature, and don't like to venture more than about 200 yards (180 m) from their home place.

To avoid the problem of having chickens forget where they live, you might construct the range shelter so that the flock is entirely confined by it. An advantage to range confinement is its superior ability to exclude predators. Its disadvantage is the need to move the shelter more often. Since the unit gives chickens less space to roam, you don't have to move it as far — only to the nearest patch of fresh pasture.

You can put young birds on range as soon as they feather out, raising them away from older birds while they develop immunities through gradual exposure to the diseases in their environment. In a warm climate, a flock can be kept on range year-round. Where winter weather turns nasty, the flock may need to be moved to permanent housing during the colder months.

Despite its many advantages, a distinct disadvantage to putting chickens out to pasture is their greater susceptibility to predators. A good fence goes a long way toward solving the predator problem.

Fencing

Whether your chickens have only a small yard or are free to roam the range, you'll need a stout fence to keep them from showing up where they aren't wanted and to protect them from predators. The fence should be at least 4-feet (120 cm) high so predators won't climb over and chickens won't fly out. It may need to be higher if you raise flyers such as Leghorns, Hamburgs, Old English, or many of the bantams.

The ideal chicken fence is made from tightly strung, small-mesh woven wire. The best fence I ever had was a 5-foot high (150 cm) chain-link fence that once came with a house I moved into. A chain-link fence isn't one I can recommend, though, because of the exorbitant cost of building one new.

The most common chicken fencing material is "chicken wire," or "poultry netting," consisting of 1-inch (2.5 cm) mesh, woven in a honeycomb pattern. Take care to specify galvanized wire designed for *outdoor* use. Chicken wire designed for indoor use rusts away all too soon when used for outdoor fencing. Many people use so-called "turkey wire" with 2-inch (50 mm) mesh because it's cheaper than chicken wire, but it doesn't hold its shape as well.

A better though more expensive option is yard-and-garden fencing with 1-inch (2.5 cm) spaces at the bottom, graduating to wider spaces toward the top (thus using less wire to keep the cost down). The smaller openings at the bottom are designed to keep small chickens from slipping out and to keep small predators from slipping in. Birds and predators can't sneak under if you pull the fence tight and attach it to firm posts that don't wobble. As further insurance, place pressure-treated boards along the ground and staple them to the bottom of the fence.

Downy chicks can pop right through most fences, but they won't stray far if they have a mother hen inside the yard to call them back. Chicks outside the fence are, however, vulnerable to passing dogs and cats. To keep them in, get a roll of 12-inch (30 cm) wide aviary netting, which looks just like chicken wire but has openings half the size. Attach the aviary netting securely along the bottom of your regular woven wire fence.

Electric Fencing

If you need additional protection from dogs and other predators, string electrified scare wires along the top and outside bottom, 8 inches (20 cm) away from your fence. The top wire keeps critters from climbing over, and the bottom wire discourages them from snooping along the fence, pushing against it, or attempting to dig under it.

If you're putting up electric wires anyway, consider building an all-electric fence. We have used electric fencing to successfully confine our chickens for many years. It's relatively inexpensive and virtually predator proof. "Virtually" doesn't mean "entirely" — even the best fence won't stop hungry hawks or opossums.

One good electrified chicken fence is made from electroplastic netting. Chickens can see the netting more easily than they can see individual horizontal electric wires, and the fence comes completely preassembled so it's easy to move when the flock needs fresh ground. Electroplastic net for poultry comes

Fence with Electrified Scare Wires

INSULATOR FOR SCARE WIRE

8" (10CM)

"YARD & GARDEN" WIRE MESH 4 FEET (120CM) OR HIGHER

8" (10CM)

*Attach wire mesh fencing to the side of the posts **away** from the chicken yard. Electrified scare wires toward the top and bottom will discourage four-legged predators.*

ALLAN DAMEROW

The electroplastic poultry netting in this photograph was made by Premier Fence Supply.

in two basic heights: the shorter version for sedate breeds is a little over 20-inches (50 cm) high; the taller version is 42-inches high (105 cm).

A controller (device transmitting electrical energy to the fence) that plugs in will give the fence more zap, especially when fast-growing weeds drain its power, but out in a field you can use a battery-operated energizer of the sort sold to control grazing livestock or to protect gardens. A lightweight net fence with a battery-powered controller lets you easily move the fence each time you move the shelter.

Sources for electroplastic netting are listed in the appendix. For full construction details on electric fencing in general, consult a comprehensive book such as *Fences for Pasture & Garden*, also listed in the appendix.

Although chickens aren't as susceptible to getting zapped as other livestock because of their small feet and protective feathers, they do learn to respect an electric fence. But first they have to know their home territory. Whenever you move chickens to a new coop, confine them inside for at least a day. When you let them into the yard, they won't stray far from home.

CHAPTER 3

FEEDING

THE MORE ATTENTION you pay to your feeding program, the greater your rewards will be in terms of tasty meat, abundant eggs, good fertility and hatchability, healthy chicks, or show awards. How you feed your chickens depends on their age, their purpose, the time of year, and the availability of forage.

Rations

The best way to make sure your chickens get a nutritionally balanced diet is to buy commercial rations. In most rations, corn supplies starch (energy) and soybean meal supplies protein. Rations for chicks contain a high amount of protein. As birds grow, they gradually need less protein and more starch.

The variety of choice you have in rations will depend a good deal on where you live. You'll have more choices if you live in an area where chicken keeping is big business. Formulations include:

- chick ration
  - medicated starter
  - non-medicated starter
- broiler ration
  - grower
  - finisher
- pullet ration
  - grower
  - developer

- lay ration
  - 16 percent protein
  - 18 percent protein
- breeder ration
- scratch grains

In many areas, developer, finisher, and breeder rations are not available, and lay ration comes only in 16 percent protein. Your choice then narrows down to two basic rations: combination chick starter/grower (either medicated or non-medicated) and lay ration (most likely 16 percent).

Chick starter is for newly hatched chicks. The medicated version is designed to prevent coccidiosis, the most common cause of death among chicks. The feeding of chicks is discussed more fully on page 232.

Broiler grower is designed for meat birds 3 to 4 weeks of age; finisher helps with their final fleshing out. The feeding of meat birds is fully discussed on page 72.

Pullet grower and developer are designed to help replacement pullets grow into top-notch layers. A complete discussion on feeding pullets appears on page 115.

A balanced lay ration for mature hens contains 16 percent protein, plus all the other nutrients layers need to keep up with egg production. In areas where temperatures soar during summer, an 18 percent formulation helps keep hens laying when hot weather causes them to eat less. The feeding of hens is discussed on page 118.

While lay ration is sufficient for egg production, it does not contain all the nutrients necessary for good hatchability. Breeder ration contains the high protein and other extra nutrients needed for the production of hatching eggs. The feeding of breeders is discussed on page 191.

Except during and immediately before breeding season, cocks do not need the same high-nutrient ration as laying hens. If you house cocks with hens, you can hardly keep them from pecking at the same trough. If you house cocks separately, you can save feed money by mixing their ration with scratch grains to reduce the total protein.

Scratch

Chickens love scratch — a mixture containing at least two kinds of grain, one of them usually cracked corn. Scratch has many functions.

- It may be used as a training device: throw down a handful while crooning "here chick, chick, chick" and your chickens will soon learn to come when you call.
- Scratch can be used to trick chickens into stirring up their coop's bedding to keep it loose and dry. Toss a handful over the litter once a day (traditionally late in the afternoon when birds are thinking of going to roost) and your chickens will scramble for it.

◆ Scratch gives range-fed layers an extra energy boost and helps to raise plump, tasty corn-fed broilers.
◆ It can also be used to reduce the protein content in a maintenance diet for cocks, thus reducing the cost of feeding them.

Too much scratch added to rations radically reduces a bird's total protein intake. In the diet of growing birds, reduced protein may lead to feather picking, which occurs when birds eat protein-rich feathers to obtain enough of that important nutrient. In the diet of laying hens, reduced protein lowers egg production and makes biddies fat and unhealthy.

Like people, chickens have preferences for the foods they eat. When you feed your flock a mixture of scratch grains, some may eat it all, some may leave the milo, and some may pick out only the corn. Avoid mixtures containing barley. Even if your chickens eat it, they'll find it hard to digest.

During winter when a flock needs more energy to stay warm, increase the scratch ration. In summer when energy needs go down, reduce the scratch or switch to whole oats. Preliminary studies at Nebraska and elsewhere show that feeding oats to hens minimizes heat stress and improves egg production in hot weather.

Molting and Protein

Since feathers are 85 percent protein, a chicken's need for dietary protein goes up when it grows new feathers during the annual molt. Commercial layers compensate for this increased need by stopping production to make dietary protein available for feather growth. Purebred hens don't always molt as quickly, and some continue to lay on a reduced basis. Supplemental protein will help them out, and will improve the feathers of show birds.

When your birds are about to molt — their plumage will take on a dull look — toss them a handful of dry cat food every other day until the molt is over. Some poultry folks use dog food, not realizing that dog food (like chicken feed) gets much of its protein from grains. Cat food, on the other hand, contains animal protein, which is rich in the amino acids a chicken needs during the molt.

Adjusting Protein

You can easily adjust the protein in your flock's diet by combining their regular ration with a supplemental ration. You'll need to know not only how much protein you want to end up with, but also the protein contents of both the regular ration and the supplement, which should be listed on their respective labels. To raise protein, choose a supplement that's higher in protein than the regular ration; to reduce protein, choose a supplement that's lower in protein than the regular ration.

Using a method called Pearson's square, you can easily determine how much ration and how much supplement you must combine to get the protein content you want. Begin by drawing a square on a piece of paper. In the upper left corner, write the percentage of protein indicated on the label of the regular ration. In the lower left corner, write the percentage of protein contained in the supplemental ration. At the center of the square, write the percentage of protein you want to end up with.

Moving from the upper left toward the lower right (following the arrow in the illustration), subtract the smaller number from the larger number. Write the answer in the lower right corner. Moving from the lower left toward the upper right (again following the arrow), subtract the smaller number from the larger number. Write the answer in the upper right corner.

The number in the upper right corner tells you how many pounds of regular ration you need. The number in the lower right corner tells you how many pounds of supplement you need.

The illustration shows two typical examples. In the first case, 16 percent lay ration is combined with 8 percent scratch to create a 9 percent cock maintenance diet. Note that, since we want to *reduce* the protein content, the number in the lower left corner must be *less* than the number in the upper left corner. Pearson's square shows that we need to combine 1 pound of lay ration with 7 pounds of scratch to get a 9 percent cock maintenance ration.

In the second example, 16 percent lay ration is combined with 31 percent

Pearson's Square

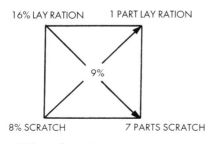

16% LAY RATION   1 PART LAY RATION

9%

8% SCRATCH   7 PARTS SCRATCH

9% cock maintenance ration

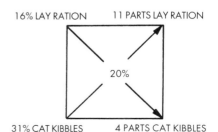

16% LAY RATION   11 PARTS LAY RATION

20%

31% CAT KIBBLES   4 PARTS CAT KIBBLES

20% breeder ration

cat kibble to create a breeder flock ration containing 20 percent protein. Since we now want to *increase* the protein content, the number in the lower left corner is *greater* than the number in the upper left corner. Pearson's square shows that we should mix 11 pounds of lay ration with 4 pounds of kibble to get a 20 percent breeder ration.

Different feedstuffs have different weights, so you won't get an accurate mix if you measure by volume (bucketfuls) instead of by weight (pounds or kilograms). If you don't have a spring scale for weighing feed, weigh yourself on a bathroom scale holding an empty bucket. Add feed to the bucket until you increase the total weight by the amount you need.

In combining feedstuffs, use rations of similar consistency. If, for example, you use soybean meal to boost the protein in pelleted ration, the meal will filter out and fall to the bottom of the trough. You'd do better to combine soybean meal with crumbles, perhaps moistening the mash at feeding time to keep the meal from sifting out.

Whenever you adjust the protein in your flock's diet by more than a percentage point or two, make the change gradually. Too rapid a change can cause stomach upset and diarrhea.

Ration Requirements

Type	Age	Ration	% Protein
Broilers	0–3 weeks	broiler starter	20–24
	3 weeks–butcher	broiler finisher	16–20
Layers	0–6 weeks	pullet starter	18–20
	6–14 weeks	pullet grower	16–18
	14–20 weeks	pullet developer	14–16
	*20 + weeks	layer	16–18
Cocks	maintenance	layer + scratch	9
Breeders	*20 + weeks	breeder	18–20

*Layer or breeder ration should not be fed to pullets until they start laying at 18–20 weeks for Leghorn-type hens, 22–24 weeks for other breeds.

Home-Mixed Rations

Mixing rations is the most complex aspect of poultry management and isn't something you should undertake if you're just starting out. Ration formulation requires:

- ◆ availability of appropriate feedstuffs
- ◆ analysis of feedstuff composition
- ◆ knowledge of the nutritional needs of chickens
- ◆ ability to mix feed in a quantity your flock will use within four weeks

If you decide to mix your own, you might purchase all the feedstuffs separately or grow some of them yourself. You'll need to include carbohydrates for energy (corn and/or other grains), protein (a combination of ingredients that together supply all the amino acids needed to create a complete protein), and vitamins and minerals (some of which will be in carbohydrate and protein ingredients; others will require additional feedstuffs).

Obtaining a nutritional analysis of each ingredient can be a daunting prospect, since the nutrient breakdown varies with an ingredient's source, the time of year it was grown, and the place where it was grown. Having small amounts of feedstuffs analyzed as you acquire them may not be economically feasible. As an alternative, you might use the averages suggested in a sourcebook such as *Nutrient Requirements of Poultry* (listed in the appendix), a volume that also discusses the nutritional needs of chickens. Additional information may be available from your state Extension office.

Depending on the volume of feed you use, you may find a local mill willing

Home-Mixed Rations

Ingredient*

Coarsely ground grain (corn, milo, barley, oats, wheat, rice, etc.)

Wheat bran, rice bran, mill feed, etc.

Soybean meal, peanut meal, cottonseed meal (low gossypol), sunflower meal, sesame meal, etc.

Meat meal, fish meal, soybean meal

Alfalfa meal (not needed for range-fed birds)

Bone meal, defluorinated dicalcium phosphate

Vitamin supplement, (supplying 200,000 I.U. vitamin A, 80,000 I.C.U. vitamin D_3, 100 mg riboflavin)

Yeast, milk powder (not needed if vitamin supplement is balanced)

Ground limestone, marble, oyster shell

Trace mineral salt or iodized salt (supplemented with 0.5 ounces manganese sulfate and 0.5 ounces zinc oxide)

　　　　Total

*If possible, use a combination of ingredients from each category.

From: "Feeding Chickens," *Suburban Rancher*, Leaflet #2919, University of California.

to mix for you. If not, you'll have to get your own mixer. You might look for an old hand-operated mixer or, if you have access to a tractor, get a mixer that operates off the PTO.

Despite its complexities, formulating your own rations has the advantage of letting you control what your chickens eat, and you *may* save money in the process. If you're interested in mixing your own rations, ask local feed outlets and your state Extension poultry specialist to help you find others who mix their own, and solicit their help in ironing out the wrinkles based on the availability of local feedstuffs.

How Much to Feed

The amount a chicken eats varies with the season and temperature, as well as with the bird's age, size, weight, and/or rate of lay. Since chickens eat to meet their energy needs, how much a bird eats depends also on the ration's energy density. Chickens fed the same ration year-round will eat more during cold weather, because they need more energy to keep their bodies warm.

A chicken that doesn't get enough to eat won't grow or lay well. A chicken may eat too little if it goes through a partial or hard molt, if it's low in peck

	Imperial			Metric	
Starter	Grower	Layer	Starter	Grower	Layer
46 lb.	50 lb.	53.5 lb.	20.5 kg	22.5 kg	24 kg
10 lb.	18 lb.	17 lb.	4.5 kg	8 kg	7.5 kg
29.5 lb.	16.5 lb.	15 lb.	13 kg	7 kg	6.7 kg
5 lb.	5 lb.	3 lb.	2.3 kg	2.3 kg	1.3 kg
4 lb.	4 lb.	4 lb.	1.8 kg	1.8 kg	1.8 kg
2 lb.	2 lb.	2 lb.	1 kg	1 kg	1 kg
+	+	+	+	+	+
2 lb.	2 lb.	2 lb.	1 kg	1 kg	1 kg
1 lb.	2 lb.	3 lb.	0.5 kg	1 kg	1.3 kg
0.5 lb.	0.5 lb.	0.5 lb.	200 g	200 g	200 g
100 lb.	100 lb.	100 lb.	45 kg	45 kg	45 kg

order, if the weather turns hot, or if it finds its ration unpalatable. Birds can be particularly fussy about texture. They don't like dusty or powdery mash. They also don't like (and shouldn't be fed) moldy or musty rations.

If your chickens don't seem to be eating enough, perk up their appetites. Begin by feeding more often, even though the trough may already be full. Offer variety — chickens are particularly fond of milk, cottage cheese, tomatoes, and salad greens, which can all be safely given in small amounts. Moistening the mash may also increase appetite: for each dozen birds, stir a little water into ¼ pound (112 g) of mash fed daily. If your appetite-stimulating attempts fail, your chickens may be in poor health (see chapter 12).

Feed that disappears too fast is a *sure* sign that something's wrong. It may mean your chickens are infested with worms. Take a sampling of droppings to your vet for a fecal test and worm your chickens if necessary. In winter, rapidly disappearing feed may mean your chickens are too cold. Eliminate indoor drafts and increase the carbohydrates in their scratch mix. Disappearing feed may not be your chicken's fault at all — make sure rodents, opossums, wild birds, and other creatures are not dipping their snouts into the trough.

Feeding Methods

Two different methods are used to feed rations to chickens — free choice feeding and restricted feeding.

Free-choice feeding involves leaving rations out at all times so chickens can eat whenever they wish. It's simpler than restricted feeding because you need to fill the trough only infrequently. We, for example, fill our tube feeders daily. Others I know fill their feeders only on weekends. In either case, the feeders hold enough to carry the flock through to the next feeding.

The obvious advantages of free-choice feeding are that it saves time and ensures that no chicken goes hungry. On the down side, feed is always available where rodents, wild birds, and other livestock (notably goats) may gobble it down; feed can get dirty, wet, or moldy between feedings; and some breeds, especially the heavier ones, tend to get fat when fed free choice. When choosing a method, you need to decide whether the advantages outweigh the disadvantages.

Restricted feeding entails feeding chickens more often but giving them only small amounts at a time. Show birds may be kept on a restricted regime so they'll look forward to human visits, in which case they're usually fed as much as they'll eat in a fifteen-minute period, twice daily.

Older lightweight hens and breeders in the dual-purpose or meat categories may be put on a restricted diet to keep them from getting fat and lazy. Pullets of the heavier breeds may be fed limited amounts to keep them from maturing too rapidly, since early laying results in fewer eggs of smaller size.

A restricted feeding program is time consuming and can cause those chickens lowest in peck order not to get enough to eat. A restricted feeding program works only if you have enough feeders so that all birds can eat at the same time. Because birds eat quickly and then have plenty of time to get bored, restricted feeding may lead to cannibalism (see page 281).

Cost Reduction

Exclusive of housing, feed accounts for 70 percent of the cost of keeping chickens, so it's only natural to look for ways to keep the cost down. One way to reduce costs is to supplement your flock's diet with table scraps such as leftover baked goods, fruit, or vegetable peelings. A few caveats:

- ◆ Don't overdo or the resulting nutritional imbalance may cause slow growth, reduced laying, and poor health.
- ◆ Don't feed raw potato peels, which chickens can't easily digest — cook potato peels or avoid them.
- ◆ Don't feed anything spoiled or rotten, which can make chickens sick.
- ◆ Don't feed strong-tasting foods like onions, garlic, or fish, which impart an unpleasant flavor to poultry meat and eggs.

If you have access to hay chaff (the bits of vegetation that collect at the bottom of a hay manger), collect it as a cheap and nutritious supplement for your flock. If you raise dairy animals and have excess milk, modest amounts make a terrific supplement. Even better is whey left over from cheese making — it contains most of the whole milk's protein and little of the fat.

Another way to reduce the cost of rations is to let your chickens graze in a pasture or on a lawn for at least part of the day, provided the grass has not been sprayed with toxins. Foraging for plants, seeds, and insects lets a flock balance its own diet while it eats less of the expensive commercial stuff, saving you 30 to 60 percent of the cost of mixed rations.

Range Feeding

Range feeding is one of many old ideas in agriculture that have come around again. Before big business took over the poultry industry, family flocks roamed the back forty, if not the front lawn. Today whole books are written on how to let your chickens eat grass.

As you might expect, short pasture perennials are more suitable for chickens than taller plants. Plants that are in the vegetative, or growing, stage are more nutritious than tough, stemmy plants. Unless your flock follows some other kind of livestock in grazing rotation, during times of rapid vegetative growth — when plants grow faster than the chickens can eat them — you may have to get out the lawn mower or BushHog to keep the pasture mowed down. Cutting plants short not only keeps them growing, but also lets in sunlight that helps reduce the build-up of infectious organisms.

Chickens tend to stay close to their housing and can quickly overgraze the area, trample the pasture, and destroy plants by digging holes for dust baths. They will be more inclined to forage widely where trees give them a sense of security (not to mention shade). Another way to encourage birds to venture forth is to space waterers and feeders at some distance from their housing.

You might also scatter scratch grains on the ground, choosing a different place each day so the chickens don't keep scratching up one area. Since foraging causes chickens to burn off extra energy, you can safely feed free-ranged chickens up to ¾ of a pound (360 g) of scratch per two dozen birds.

When the pasture has been grazed down, or if bare spots appear, move the chickens to a new range (by moving their range shelter to new ground or by letting them into a new yard from permanent or semi-portable housing). You'll be doing nothing more than imitating the natural conditions under which plants evolved and under which they grow best. In nature, a flock moves together to avoid predation, quickly grazes down an area (during which it scratches up the ground and deposits large amounts of manure), then moves on.

How long a flock takes to graze down a given area depends on a number of factors including the size of the flock, the kind and condition of the pasture, temperature, and rainfall. When sun, rain, and warm weather combine to help plants grow quickly, a flock might graze a given area for two weeks or more. In cool, hot, or dry weather, when plants grow slowly, a flock may graze pasture down to nothing in a matter of days. Chickens that are confined within a range house need to be moved to new ground daily.

How small an area you can confine your flock to, and how long you can keep them there, can be determined only through watchful experimentation within these parameters:

- Let chickens in when plants are no more than 5-inches (12.5 cm) tall.
- Move chickens out when plants have been grazed down to 1 inch (2.5 cm) or when bare spots appear.
- The stocking rate, or number of chickens one acre can support in a season, for a free-ranged flock on well-managed pasture is 200 birds per acre (500 per hectare).

◆ The stocking rate for a confined flock on well-managed pasture is 500 birds per acre (1250 per hectare).

◆ The longer a range shelter stays in one place, the more time it takes for the pasture to be restored once the shelter is moved.

Over the years, pasture soil will increase in acidity. When soil pH drops below 5.5 as determined by a soil test, spread lime at the rate of 2 tons per acre (4.5 metric tons per hectare). Then let the pasture rest to give plants time to rejuvenate and to break the cycle of parasitic worms and infectious diseases. For

Moving a Range Shelter

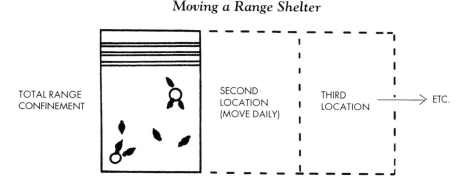

For total range confinement, move the shelter daily to a new location.

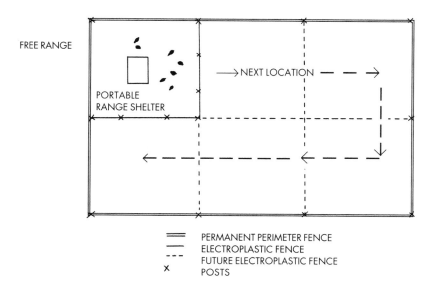

For a free-ranged flock, move the shelter when pasture has been grazed down to one inch (2.5 cm) or when bare spots appear.

a complete discussion on pasture management, consult Bill Murphy's comprehensive book *Greener Pastures on Your Side of the Fence* (listed in the appendix).

Supplements

Depending on how they are managed, chickens may need supplemental grit, calcium, phosphorus, or salt.

Grit, in the form of pebbles and large grains of sand pecked by a chicken, lodges in the bird's gizzard — an organ consisting of a tough membrane surrounded by strong muscles. When grains and fibrous vegetation pass through the gizzard, muscular action grinds them into the grit to break them up. Grit serves as a chicken's teeth.

Chickens that eat only commercially prepared mash or pellets need no grit, since the rations are sufficiently softened by the bird's saliva. Range-fed flocks need grit to grind up plant matter. Any chicken that eats grains needs grit.

Yarded or pastured birds pick up natural grit from the soil but may not get enough. Granite grit, available from any feed store carrying poultry rations, should be offered in a separate hopper as a free-choice supplement.

Calcium is needed by laying hens to keep eggshells strong. The amount of calcium a hen needs varies with her age, diet, and state of health. Older hens need more calcium than younger hens. Some illnesses cause a calcium imbalance. In warm weather, when all chickens eat less, the calcium in a hen's ration may not be enough to meet her needs. A hen that gets too little calcium lays thin-shelled eggs.

Eggshells consist primarily of calcium carbonate, the same material found in oyster shells and limestone. All laying hens should have access to a separate hopper full of ground oyster shells or ground limestone (not "dolomitic" limestone, which can be detrimental to egg production).

Phosphorus and calcium are interrelated — a hen's body needs one in order to metabolize the other. Range-fed hens obtain some phosphorus and calcium by eating beetles and other hard-shelled bugs, but they may not get enough. Phosphorus in the form of defluorinated rock phosphate or phosphorus-16 should therefore be offered to free-range hens in a separate hopper on a free-choice basis. Hens fed nothing but a commercial lay ration should get enough phosphorus without supplementation.

Salt is needed by all chickens, but only in minute amounts. Commercially prepared rations contain all the salt a flock needs. Range-fed chickens that eat primarily plants and grain may need a salt supplement. Salt deficiency causes hens to lay fewer, smaller eggs and causes any chicken to become cannibalistic.

Feeding Station

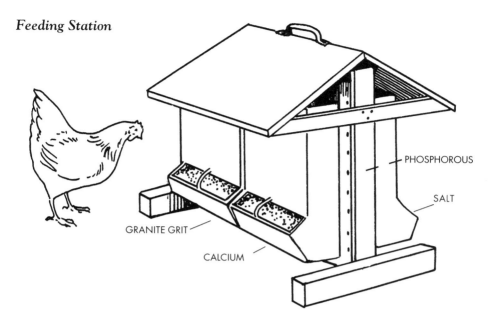

PHOSPHOROUS

SALT

GRANITE GRIT

CALCIUM

In this feeding station for dietary supplements, two hoppers face each side. Hoppers are hung for easy removal and cleaning.

Loose salt (*not* rock salt) should be available to range-fed chickens in a separate hopper on a free-choice basis.

Caution: A normal amount of salt can cause poisoning if chickens do not have access to water at all times. In warm weather when chickens need more water than usual, make sure they don't run out; in winter, do whatever it takes to keep drinking water from freezing or remove the salt hopper until the weather warms.

Feeders

Feeders come in many different styles — the two most common being a long trough and a hanging tube. Regardless of its design, a good feeder has these important features:

- ◆ discourages billing out
- ◆ prevents contamination with droppings
- ◆ is easy to clean
- ◆ doesn't allow feed to get wet

Chickens are notorious feed wasters. Feeders that encourage wastage are narrow or shallow and/or lack a lip that prevents chickens from billing out — using their beaks to scoop feed onto the ground. A feeder with a rolled or bent-in edge reduces billing out. To further discourage billing out, raise the feeder to the height of the chickens' backs. The best way to keep a feeder at the right height as a flock grows is to hang it from the ceiling by chains.

If you use a trough feeder, never fill it more than two-thirds full. Chickens waste approximately 30 percent of the feed in a full trough, 10 percent in a two-thirds full trough, 3 percent in a half-full trough, and approximately 1 percent in a trough that's only one-third full. Obviously, you'll save a lot of money by using more troughs so you can put less feed in each one.

A good feeder discourages chickens from roosting on top and contaminating feed with droppings. A trough mounted on a wall allows little room for roosting. A free-standing trough may be fitted with an anti-roosting device that turns and dumps a chicken trying to perch on it. A tube feeder should be fitted with a sloped cover to prevent roosting; unfortunately, most tube feeders don't come with covers anymore, but you can fashion one from the lid of a 5-gallon plastic bucket. It may not keep chickens from roosting, but it will keep their droppings out of the feed.

Since you fill a trough from the top and chickens eat from the top, trough feeders tend to collect stale or wet feed at the bottom. Never add fresh feed on top of feed already in the trough. Instead, rake or push the old feed to one side, and at least once a week empty and scrub the trough.

After having used trough feeders for years, I much prefer tube feeders. Since you pour feed into the top and chickens eat from the bottom, feed doesn't sit around getting stale. A tube feeder is fine for pellets or crumbles, but works well for mash only if you fill it no more

Feeders

One hanging feeder is enough for up to 30 chickens.

This trough has adjustable-height legs and an anti-roosting reel that rotates and dumps any bird that tries to hop on. Allow 4" (10 cm) of trough space for each bird, counting both sides if birds can eat from either side.

Homemade Tube Feeder

Make an inexpensive tube feeder from a 2-foot (60 cm) length of 8-inch (20 cm) stovepipe, a 15-inch (approximately 37.5 cm) hubcap from a salvage yard, a large turnbuckle from the hardware store, and a bit of wire. Drill holes into opposite sides of the stovepipe, 6 inches (15 cm) from the top, and insert the turnbuckle. Thread the eyes at the ends of the turnbuckle with chains or wire from which to hang the feeder.

At the bottom end, suspend the hubcap with three short pieces of wire, leaving a $1/2$-inch (12.5 mm) gap between the pipe and the hubcap. The hubcap's lip will keep the chickens from billing out and wasting feed. This feeder (inspired by a design in *Progressive Farmer* magazine) doesn't need a lid, since it will sway when a chicken lands on it, discouraging the chicken from perching.

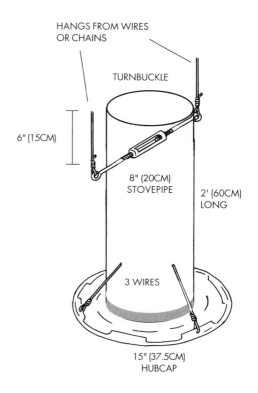

Homemade Hanging Feeder

than two-thirds full. Otherwise the mash may pack and bridge, or remain suspended in the tube instead of dropping down.

Placing feeders inside the coop keeps feed from getting wet, but encourages chickens to spend too much time indoors. Hanging feeders under a covered outdoor area is ideal for keeping feed out of the rain and for encouraging the flock to spend more time in fresh air. If you have to keep feeders indoors, for good litter management move them every two or three days to prevent concentrated activity in one area.

If you feed free choice, put out enough feeders so at least one-third of the flock can eat at the same time. If you feed on a restricted basis, you'll need enough feeders so the whole flock can eat at once.

Feed Storage

It's a good idea to keep extra feed on hand so you won't run out, but don't stock up too far ahead. From the moment it's mixed, feed starts losing nutritional value through oxidation and other aging processes. Any feed should therefore be used within about four weeks of being milled. Allowing a week or two for transport and storage at the farm store, buy only as much as you can use in two to three weeks.

Store feed off the floor on pallets or scrap lumber, away from moisture. After opening a bag, pour the feed into a clean plastic trash container with a tight-fitting lid. A plastic container is preferable to a galvanized can, since metal sweats in warm weather, causing feed to get wet and turn moldy. A closed container will slow the rate at which feed goes stale and will keep out rodents. A 10-gallon container will hold 50 pounds of feed. By happy coincidence, feed comes in 50-pound (or sometimes 25 pound) bags (in metric: 25 or 40 kg bags).

Keep the container in a cool, dry place, out of the sun. Never pour fresh feed on top of old feed. Use up all the feed in the container before opening another bag. If you have a little feed left from a previous batch, pour it into the container's lid, pour the fresh bag into the container, and put the older feed on top where you'll use it first.

Feed spilled around the container will attract rodents, and they can cost you money by eating up incredible quantities of feed. They can also carry diseases that may infect your birds. For a discussion on rodent-control measures, see page 264.

Water

A chicken drinks often throughout the day, sipping a little each time. A chicken's body contains more than 50 percent water, and an egg is 65 percent water. A bird, therefore, needs access to fresh drinking water at all times in order for its body to function properly. A hen that is deprived of water for 24 hours may take another 24 hours to recover. A hen deprived of water for 36 hours may go into a molt, followed by a long period of poor laying from which she may never recover.

Depending on the weather and on the bird's size, each chicken drinks between 1 and 2 cups (237–474 ml) of water each day. Layers drink twice as much as nonlayers. In warm weather, a chicken may drink two to four times more than usual. When a flock's water needs go up during warm weather and the water supply remains the same, water deprivation can result. Water deprivation can also occur in winter if the water supply freezes.

Even when there's plenty of water, chickens can be deprived if they simply don't like the taste. Medications, for example, can cause chickens not to drink. Do not medicate water when chickens are under high stress, such as during hot weather or during a show.

Large amounts of dissolved minerals can make water taste unpleasant to chickens. If you suspect your water supply contains a high concentration of minerals, have the water tested. If total dissolved solids exceed 1,000 parts per million (ppm), look for an alternative source of water for your flock.

Chickens prefer water at temperatures between 50°F and 55°F (10–13°C). The warmer the water, the less they'll drink. In summer, put out extra waterers and keep them in the shade, and/or bring your flock fresh, cool water often. In cold weather, make sure water doesn't freeze: bring your flock warm water at least twice a day (but avoid increasing coop humidity by filling indoor fountains with steaming water), use an immersion heater in water troughs, place metal fountains on pan heaters, and wrap heating coils around automatic watering pipes. Water-warming devices are available through farm stores and livestock supply catalogs.

Water Heaters

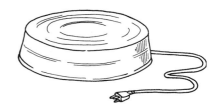

To keep water from freezing indoors, set metal fount on thermostatically controlled heating pan

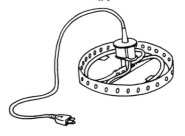

Outdoors, drop sinking heater into bucket.

Drinking Water Quality

Content	Maximum Recommended Level
Iron	2 ppm
Nitrates	45 ppm
Sulfates	250 ppm
Sodium chloride (salt)	500 ppm
Total dissolved solids	1,000 ppm
Total alkalinity	400 ppm
pH	8.0

Source: *Cobb Broiler Manual*, Cobb, Inc.

Waterers

Chickens should not have to get their drinking water from puddles or other stagnant unhealthful sources but should be provided fresh, clean water in suitable containers. Waterers, like feeders, come in many different styles. The best drinkers have these features in common:

- ◆ They hold enough to water a flock for an entire day.
- ◆ They keep water clean and free of droppings.
- ◆ They don't leak or drip.

Automatic, or piped-in, water is the best kind because it ensures that a flock never runs out. But piped-in water isn't without disadvantages. Aside from the expense of running plumbing to the chicken coop, water pipes can leak if not properly installed and freeze in winter unless buried below the frost line or wrapped in electric heating tape. Automatic drinkers can also get clogged, and so must be checked at least daily.

Properly managed piped-in water is handy for birds kept in cages, since it saves the trouble of having to distribute water to each cage by hand. Automatic devices come in two basic designs, nipples and cups.

Nipples dispense water when manipulated by each individual bird. Since birds have to learn how to drink from a nipple, you'll need to spend time watching to make sure all your birds know how to drink and to help those having trouble. One nipple can serve up to five birds.

Cups hold a small amount of water, the level of which is controlled by a valve that releases water each time a bird drinks. Cups come in small and large sizes. The smaller size is for birds in cages. The larger size can be used by a flock. Provide one large cup for up to 100 birds.

Inexpensive plastic 1-gallon drinkers are fine for young birds, but they don't hold enough for many older birds (and usually get knocked over by rambunctious adults). In addition, plastic cracks after a time, and the cost of replacing those inexpensive waterers adds up fast.

Metal waterers are sturdier than plastic and come in larger sizes holding 3 gallons, 5 gallons, or more. As in all things, you get what you pay for — a cheap metal waterer will rust through faster than a quality drinker. Whether you use plastic or metal, set the container on a level surface so the water won't drip out. To help birds drink and to reduce litter contamination, the top edge of the waterer should be approximately the height of the birds' backs.

Placing the container over a droppings pit confines spills so chickens can't walk or peck in moist, unhealthful soil. Build a wooden frame of ½ x 12 x 42 inch (12.5 x 304 x 1066 mm) boards. Staple strong wire mesh to one side and set the box, wire side up, on a bed of sand or gravel. Place the waterer on top so chickens have to hop up onto the mesh to get a drink.

Provide enough waterers so at least one-third of your birds can drink at the same time. No less than once a week, clean and disinfect waterers with a solution of chlorine bleach.

Automatic Watering for Caged Chickens

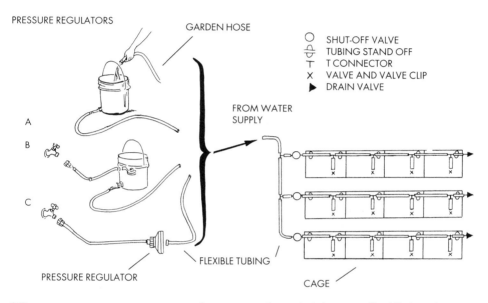

Water enters automatic waterers three ways: through (a) manually filled tank, (b) float valve–regulated tank, or (c) pressure-regulated direct water supply.

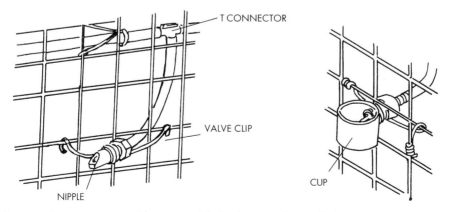

Automatic waterers provide cages with fresh water from nipples or cups.

Homemade Waterer

Make an inexpensive 1-gallon metal waterer from an empty number 10 can, available for the asking from many cafeterias and restaurants. For the base, you'll need a round cake pan, 2 inches (5 cm)wider than the can.

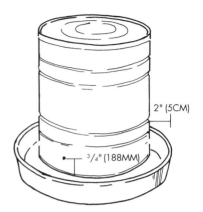

Punch or drill two holes in the can, opposite each other and ¾ inch (188 mm) from the open end. Fill the can with water, cover it with the upside-down cake pan, and flip the whole thing over. The little holes let water dribble out every time a chicken takes a sip, keeping the pan filled with fresh water.

CHAPTER 4

MEAT BIRD MANAGEMENT

OF ALL THE DIFFERENT FORMS of livestock, chickens give you meat on the table with the least amount of time and effort. In a matter of weeks your chicken-keeping chores are over, and your freezer is full of tasty, healthful poultry.

Meat Breeds

Your first decision in establishing a meat flock is what kind of birds to raise. You have two basic choices:

- ◆ a commercially developed Cornish-cross strain
- ◆ an old-fashioned heavy or dual-purpose utility breed.

Cornish-cross broilers have the advantage of being interested in only one thing — eating. Since all they do is eat, they grow fast and tender. But this characteristic also makes them coop potatoes. When they're not eating, they have nothing to do except sit around developing bad habits like cannibalism or getting sick and dying. Managing hybrids therefore requires careful attention.

Old-fashioned utility purebreds are hardier than commercial meat strains, but they grow more slowly. Where a meat hybrid reaches 5 to 6 pounds (2–3 kg) in 8 weeks or less, a purebred takes 13 weeks or more to reach the same weight. Compared to a commercial meat strain, the purebred utility strain will be lower in fat and firmer in texture, and will have a stronger chicken flavor due to its older age. According to some reports, the longer growing period of purebred utility strains also makes them more nutritious, because they have more time to develop complex amino acids.

Some people describe nonhybrid meat as "tough" and suitable only for slow, moist cooking. I couldn't agree less. After years of enjoying the meat of

65

dual-purpose birds cooked in all the same ways as store-bought meat, I find the latter has a bland taste and an unnatural mushy texture.

We raise meat birds as an adjunct to our laying flock. Each spring we hatch a batch of dual-purpose chicks to get replacement pullets. When the pullets are big enough to go out on range, we separate them and confine the surplus cockerels until they reach fryer size. Later, any pullets that don't measure up are culled as roasters. When the remaining pullets come into lay, our old hens go into the freezer as stewing hens.

The idea of raising meat birds as a by-product of the laying flock is far from new. It is, in fact, how today's broiler industry got started in the first place. In the 1920s, many housewives like Mrs. Wilmer Steele of Ocean View, Delaware, purchased chicks every year to raise as layer replacements. One year Mrs. Steele got 500 chicks instead of the 50 she had ordered, so she raised the surplus for meat and sold them at a dandy profit. The next year she deliberately bought 1,000 chicks and again the money rolled in. When his wife worked her way up to 25,000 chicks a year, Mr. Steele retired from the Coast Guard to stay home and help. For years thereafter, Delaware was the center of the broiler industry and development of efficient meat strains.

Meat Classes

The class of poultry meat you prefer may influence your choice between hybrids and nonhybrids. Meat birds are divided into six basic classes:

Rock-Cornish game hen — Not a game bird and not necessarily a hen, but a Cornish, Rock-Cornish, or any Cornish-cross bird, usually 5 to 6 weeks old, coming in two size ranges: less than 1 pound (0.45 kg) and about 1½ pounds (0.68 kg). To get well-rounded "game hens," you have to start with hybrid chicks.

Broiler, or *fryer* — A young tender bird, usually weighing 4 to 4½ pounds (1.8–2 kg) live weight, less than 13 weeks of age, and of either sex. Has soft, pliable, smooth-textured skin and a flexible breastbone, suitable for almost any kind of cooking. Hybrids and nonhybrids alike make good broiler/fryers.

Roaster — A young tender bird, usually weighing 6 to 8 pounds (2.7–3.6 kg) live weight, 3 to 5 months of age, and of either sex. Has soft, pliable, smooth-textured skin and a breastbone that's less flexible than a broiler/fryer, suitable for roasting whole. Roasters may be hybrid or nonhybrid, but because of the former's voracious appetite beyond the fryer stage, the latter are more economical to raise as roasters.

Capon — A young desexed male chicken, usually under 8 months old, whose meat is tender and skin is soft, pliable, and smooth-textured, suitable for roasting whole. Since hybrids are specialized for early rapid growth, caponized birds are more often nonhybrid.

Stewing hen, baking hen, or *"fowl"* — A mature (usually 10 months or older) female chicken with less tender meat and a nonflexible breastbone, requiring moist cooking methods such as stewing, braising, or pressure cooking. Stewing hens are laying hens that are no longer economically productive.

Cock, or *rooster* — Any male chicken that has entered the stag stage, when combs and spurs develop and meat turns dark, tough, strong tasting, and no longer fit to eat.

Color Preferences

Many fine meat breeds never caught on in the United States simply because of the color of their skin. The skin color of a meat bird is a matter of preference, and consumers generally prefer what they've been taught to like. As a general rule, European consumers favor white-skinned breeds, Orientals like black-skinned breeds, and Americans prefer yellow-skinned breeds. All of the hybrids developed for meat production in this country have yellow skin.

In North America, birds with white plumage are preferred for meat because they look cleaner when plucked than dark-feathered birds. Regardless of feather or skin color, the taste is pretty much the same. Since many people now remove the skin before cooking or serving chicken, preferences in feather and skin color are less important than they once were.

Managing Meat Birds

Your decision regarding what kind of birds to raise will, to some extent, be determined by how you wish to raise them. Methods for raising broilers fall into three basic categories:

- ◆ indoor confinement
- ◆ range confinement
- ◆ free range

Skin Color of Meat Breeds

Yellow

Brahma
Cornish
Dominique
Jersey Giant
Langshan
New Hampshire
Plymouth Rock
Rhode Island
Red
Wyandotte

White

Australaorp
Dorking
Houdan
Orpington
Sussex

Pinkish

Catalana

Black

Silkie

The last two methods are sometimes grouped together as "range feeding," a concept favored by people who prefer their food to be produced as naturally as possible. Since the first two methods involve confinement and the last two involve pasturing, the middle method, range confinement, bridges the gap between indoor confinement and free range.

Indoor Confinement

Indoor confinement is the preference of large commercial growers because it lets them maximize capacity in terms of capital investment and facilities. It is also practiced by small-flock owners who don't have much space for raising meat birds. It involves housing chickens on litter and bringing them everything they eat.

On the up side, indoor confinement requires less land than the other two methods (all you need is a building) and less time (set up your facilities properly, and you should spend no more than a few minutes each day feeding and watering). On the down side, feed costs are higher than for range methods.

The goal of indoor confinement is to get the most meat for the least cost by efficiently converting feed into meat. The standard feed conversion ratio is 2:1 — each bird averages 2 pounds (kilograms) of feed for every 1 pound (kilogram) of weight it gains. To get a feed conversion ratio that good, you must raise a commercial meat strain developed for its distinct ability to eat and grow.

On a commercial scale, feed conversion efficiency is improved by adding antibiotics to rations. No one is quite sure why it works, but some researchers speculate that antibiotics thin a bird's intestinal walls and thereby improve nutrient absorption. While no detectible drug residues are legally allowed in commercially produced meat, in practice no one seems to be minding the store. Antibiotic residues in meat are harmful to humans for several reasons: they disturb the natural balance of microflora in our intestines, they can induce a serious reaction in those of us who are allergic to antibiotics, and they cause resistance to prescription drugs.

Not all commercial producers use growth promoters. Like most small-scale flock owners, some producers rely on careful management and good sanitation for efficient feed conversion. If one of the reasons you raise your own chickens is to get drug-free meat, avoid feeds containing antibiotics and other medications.

Even if your management is meticulous, you can't raise broilers indoors for the same low cost you would pay at the grocery store. For starters, the cost of chicks by the dozen is much higher than the cost of buying by the thousands. The same holds true for buying feed by the bagful, instead of periodically having a truck roll in to fill your silo.

You won't get the same high feed conversion ratio, either, unless your facilities are designed to encourage feed consumption, right down to conveyor belts that keep feed moving to attract birds to peck. On a small scale, the best you can do to pique interest in eating is to feed birds often to stimulate their appetites.

Efficient feed conversion also means keeping housing at a temperature between 65°F and 85°F (18°C–29°C), which entails providing supplemental heat or crowding birds enough for them to keep each other warm. If you give your meat birds more room than the minimums shown in the accompanying chart, be prepared either to heat their housing or feed them longer to get them up to weight.

Minimum Space for Confined Meat Birds

Age	Floor Space/Chick	
0–2 weeks	0.5 sq ft	468 sq cm
2–8 weeks	1 sq ft	936 sq cm
8+ weeks	2–3 sq ft	1872–2808 sq cm

Another aspect of efficient feed conversion is controlling light. Compared to artificial light, natural light causes birds to be more active. As a result, they use up more calories and grow more slowly. Because they have little else to do, when they're not eating they get bored and resort to feather picking and cannibalism. If you let your birds enjoy natural light, you'll have fewer problems if you limit it to no more than 10 hours a day.

During the rest of the day. provide just enough light to let birds find the feeders, but not enough to inspire them to engage in other activities. Get chicks started eating and drinking under 60-watt bulbs, placed in reflectors 7 to 8 feet (210–240 cm) above the floor. After 2 weeks, switch to 15-watt bulbs. Allow one bulb-watt per 8 square feet (624 sq cm) of living space.

The number of light hours meat birds should get per day is a matter of debate. The trend in commercial production is to shorten daylight hours for chicks 2 to 14 days old. Shorter days give them less time to eat, slowing their growth rate and thereby reducing leg problems and other complications resulting from too-rapid growth. After the birds reach two weeks of age, light hours are increased to encourage them to eat and grow.

Raising broilers under continuous light is a bad idea, in any case, since they may panic, pile up, and smother if the power fails. To get the birds used to lights-out, turn lights off at least one hour during the night. Some growers contend that as little as 14 hours of light per day is enough for efficient growth. During hot weather, when chickens get lethargic, turning lights on in the morning and evening encourages them to eat while the temperature is lower. Putting lights on a timer will save you the trouble of having to remember to flick the switch.

Since rapid growth characteristically causes weak legs, don't provide roosts for your meat birds. Leg injuries can occur when heavy birds jump down from roosts. Perching can also cause crooked keels and breast blisters. Injuries and blisters may also occur when heavy birds are housed on wire, wood, bare concrete, or packed, damp litter. Avoid these problems by housing confined meat birds on a soft bed of deep, dry litter.

Range Confinement

Like indoor confinement, range confinement involves keeping broilers in a building, but this building is portable, is kept on pasture, and is moved daily. Pasture confinement is suitable for hybrid and purebred strains alike, although (as with indoor confinement) the former will grow more quickly than the latter.

On the up side, range confinement reduces feed costs, especially if you move housing first thing each day to encourage hungry birds to forage for an hour before feeding them their morning rations. On the down side, you need enough good pasture (or unsprayed lawn) to move the shelter daily and, without fail, you must take a few minutes each day to do so. As they reach harvest size, birds will graze plants faster and, as a result, deposit a greater concentration of droppings, so they'll have to be moved more often — at least twice a day.

After the first few moves, the birds will learn to walk along as you move their floorless shelter. Don't be tempted to help things along by installing a wire floor so you can lift the birds right along with the shelter. We tried that and found that the extra weight of the birds makes the whole shebang heavy and unwieldy. Not only that but while the shelter is in motion, the birds stabilize themselves by curling their toes around the wire, causing toes to get smashed when the shelter is set back down.

The pioneers of modern small-scale commercial range confinement are Joel and Theresa Salatin of Swoope, Virginia, who describe their method in detail in their book *Pastured Poultry Profit$* (listed in the appendix). The Salatins confine up to 100 broilers per 10 x 12 x 2 foot (300 x 360 x 60 cm) pen made of chicken wire stapled to a pressure-treated wood frame and roofed with corrugated aluminum. Weather permitting, chicks may be moved from the brooder to the pen once they reach 2 weeks of age. In order for chicks to do well on pasture, they must begin foraging by the age of 25 days.

Using a homemade dolly, the Salatins move each pen daily, requiring a total of 5,000 square feet (450 sq m) of good grazing per pen per 40-day growing period. The couple raises a commercial strain that reaches a butchering weight of 4 to 4½ pounds (1.8–2 kg) in 8 weeks, the same as they would if confined indoors. The management differences are basically twofold: feed costs are less, and because the birds spend so much time grazing, they have less time to pick at each other.

Free Range

Free range is similar to range confinement, except that birds are allowed to freely come and go from their shelter. Like range confinement, free ranging reduces the cost of mixed rations. Unlike range confinement, it lets birds exercise. The extra activity creates darker, firmer, more flavorful meat, but also causes birds to eat more and grow more slowly — they may not reach fryer size until about 13 weeks.

Free ranging involves less labor than range confinement (because you don't have to move the shelter daily) but more labor than indoor confinement (because you do have to move the shelter occasionally). Compared to either form of confinement, free ranging requires more land — enough for the shelter itself as well as pasture for grazing (and trampling), multiplied several times to allow for fresh forage. You'll need at least ¼ acre (about 1,000 sq m) for 100 birds.

Utility breeds take to grazing quite readily, since they retain some of the foraging instincts of their forebears. Commercial strains don't think too much of getting out and around, but they will roam more than they do in confinement, which slows their growth and makes them less prone to leg problems.

If you raise straight-run meat birds, you'll have to separate the cockerels as soon as they become sexually active, otherwise they'll harass the pullets and neither will grow well. Sexual harassment is not a problem with confined broilers, since they go into the freezer before they begin noticing the opposite sex.

Because not everyone is willing to raise broilers an extra 5 weeks, and not everyone appreciates the full flavor and firm texture of naturally grown chicken, free ranging is used less often for meat birds than for laying hens.

Combination Management

Combination management involves raising birds indoors for 8 weeks, butchering some of them as fryers, and putting the rest on pasture for another 4 to 5 weeks to raise as roasters. This is the system that — through no deliberate design — we use for our surplus-cockerel fryers and cull-pullet roasters. Since feed conversion efficiency goes down as birds get older, raising roasters this way costs less than raising them entirely indoors, but the birds take longer to get up to roaster weight.

Melodious Growth Booster

Fowls favor euphonious sounds — a phenomenon discovered by animal physiologist Gadi Gvaryahu, who found a flock of chickens crowded around a radio inadvertently left running. As Gvaryahu discovered, you can boost the growth rate of meat birds by letting them listen to classical music.

Feeding Meat Birds

A wide variety of meticulously formulated starter, grower, and finisher rations has been developed with one thing in mind — to keep feed costs down. Newly hatched chicks need a lot of protein. As chicks grow, their protein needs go down and their carbohydrate needs go up. Since protein sources (legumes and meat scraps) are more expensive than carbohydrate sources (starchy grains), switching to rations with progressively less protein saves money.

To precisely target the protein versus energy needs of meat birds according to their stage of growth, big-time growers formulate their own rations and have them privately milled. We small-flock owners are at the mercy of local feed mills. Depending on where you live, you may have little choice in the available combination of starter and finisher or starter/grower and finisher.

How much that matters to you depends on your method of management. A confinement-fed broiler eats approximately 2 pounds of feed for every pound of weight it gains. If you raise your birds to 4 pounds, each one will gobble up at least 8 pounds of feed during its lifetime. A purebred strain may eat twice that much, a factor you can mitigate by letting your birds forage for much of their sustenance.

In any case, if all you have available to you is one all-purpose starter/grower ration, there's nothing inherently wrong with using it from start to finish. But don't expect the same rapid growth or low feed-to-meat ratio you would get with a more targeted ration. If you do have a wider choice, follow directions on the label regarding when to switch from one ration to another. Each manufacturer's feeding schedule is based on the formulations of its particular rations.

Unfortunately, standard commercial rations may not contain sufficient nutrients to sustain the rapid growth rate of broiler-cross chicks, and as a result they develop leg problems. If you raise a commercial broiler strain, supplement the rations with a vitamin/mineral mix, added either to feed or to drinking water (according to directions on the label).

The older a chicken is, the less efficient it becomes at converting feed into meat and the costlier it becomes to raise. The conversion ratio starts out below 1 in newly hatched chicks and reaches 2:1 at about the fifth or sixth week. During the seventh or eighth week, the cumulative, or average, ratio reaches 2:1 — the point of diminishing returns.

From then on, the cumulative ratio has nowhere to go but up and the amount of feed eaten (in terms of cost) can't be justified by the amount of weight gained. Although the most economical meat comes from birds weighing 2½ to 3½ pounds (1–1.5 kg), most folks prefer meatier broiler/fryers in the 4 to 4½ pound (1.8–2 kg) range. If you want nice plump roasters, be prepared to pay more per pound to raise them.

Feed Consumption Guidelines

The bigger a bird is, the more it eats. It stands to reason, then, that a flock's feed use should steadily increase as the birds grow. To estimate the minimum amount of feed 100 confinement-fed chicks should eat each day, double their age in weeks. For example, at 4 weeks old, 100 broilers should eat no less than 8 pounds of feed per day. (In metric, the age of the chicks in weeks roughly equals the minimum amount of feed, in kilograms, 100 chicks should eat each day). If feed use levels off or drops below this guideline, look for management or disease problems.

Corn-Fed Fryers

Even if you don't have available pasture, you can still enjoy some of the cost savings and fine flavor of range-fed meat by raising corn-fed fryers. These birds obtain up to 70 percent of their diet from corn or, more precisely, scratch grains. As a result, they enjoy a slower growth rate that compares with the growth of free-ranged chickens. Like ranged birds, they aren't ready for butchering until about 13 weeks.

Corn-fed fryers aren't politically correct among cholesterol-conscious nutritionists, since a diet that's heavy in grains makes chickens fat. But they are trendy in natural-food circles where the goal is to avoid feed additives by using so-called "organic" grains. The problem is, organically grown grains are hard to come by. Chances are good that the scratch you feed your chickens is the same stuff that's used, in ground-up form, to make commercial rations.

If finisher ration is available, feed chicks commercial rations for the first six weeks, switch entirely to scratch grains until the last two weeks, then let finisher supply 30 percent of their diet. If you have grower ration available, or only one starter/grower formula, feed your chicks commercial rations for the first four weeks, then switch to grower (or stay with the starter/grower) for 30 percent of their diet and scratch grains for the remaining 70 percent right to the end. In either case, as soon as you offer your birds scratch, they'll need free-choice granite grit in order to digest the grains.

Avoiding Drug Residues

To avoid drug residues in your home-grown meat, shun medicated rations and seek out sources of nonmedicated rations. Medications include low levels of antibiotics to improve feed conversion and coccidiostats to prevent coccidiosis, an intestinal disease that interferes with nutrient absorption and drastically reduces the growth rate of infected birds. Unfortunately, if you raise chicks in a warm, humid climate or at a warm, humid time of year, you can't avoid using a coccidiostat. Without it, you'll be fighting a losing battle.

Whether you use a medicated ration because you need to or because you can't find anything else, you still must find a nonmedicated feed to use during the drug's withdrawal period. The withdrawal period supposedly represents the minimum number of days that must pass from the time drug use stops until drug residues no longer show up in the birds' meat. If the label does not specify a withdrawal period, ask your feed dealer to look it up for you in his spec book. If nonmedicated rations are not available, scratch grains may be your only option during the drug withdrawal period.

If you use nonmedicated feed throughout the growing period, you will have to be especially careful to prevent coccidiosis. Keep litter clean and dry for indoor birds. Move range-fed birds as often as necessary to prevent a build-up of droppings.

In both cases, keep drinking water clean of droppings. Clean water serves more purposes that preventing disease — broilers that don't have free access to fresh water eat less and, as a result, grow at a slower rate.

Feeder Space

Provide enough feeder space so broiler chicks can eat at will, and so lower ones in the peck order won't get pushed away from feed by dominant birds. The general rule is to provide enough feeder space so at least one-third of your birds can eat at the same time. If you use hanging tube feeders, you'll need one for every 3 dozen birds. The accompanying chart shows space recommendations for trough feeders.

Trough Space per Bird

0–2 weeks	1 in.	2.5 cm
3–6 weeks	2 in.	5.0 cm
7–12 weeks	3 in.	7.5 cm
12+ weeks	4 in.	10.0 cm

Capons

Before the development of highly specialized broiler strains, meat birds were nothing more than cockerels culled from straight-run chicks raised to get replacement pullets. Those of us who keep dual-purpose flocks still raise meat birds the old-fashioned way. Due to the slow growth of these nonhybrids, we have to take care to butcher cockerels before they reach the stag stage, when their spurs and combs begin to develop and their meat toughens.

If you wish to raise meat birds beyond the stag age, cockerels must be caponized, or have their testicles surgically removed to channel their energy into continued growth rather than sexual maturity. A capon grows to the size of a small turkey — offering an alternative to the holiday gobbler — but is easier to grow (turkeys being notorious for sitting around thinking up ways to die).

The hackle, saddle, and tail feathers of a capon grow longer than those of a cockerel, but instead of developing a large comb, the capon keeps the cockerel's small, pale head. The capon has a calmer disposition than a cockerel and doesn't crow. He'll gain weight at about the same rate as a cockerel until approximately 18 weeks of age, when his growth rate surpasses that of a cockerel. The capon is more expensive to raise, though, since he isn't properly grown and finished until he's at least 20 weeks old.

Dual-purpose cockerels are separated from pullets and caponized at 5 to 6 weeks of age. Although hybrids are rarely caponized, there's no reason you can't caponize a few to raise nice plump roasters. Meat strains are caponized at 2 to 3 weeks — beyond that age, surgery becomes difficult due to the hybrid's size.

Compared to an intact cockerel (left), a capon remains calmer, grows larger, and rarely crows.

Caponizing

Before you begin, be sure to check with your local authorities about the legality of caponization in your area. In some places, the operation can only be performed by a professional veterinarian.

On the day before caponizing is scheduled, withdraw feed and water for 24 hours so full intestines won't interfere with surgery. House the birds in a wire or slat coop, where they can't eat litter.

Operate in good light. You'll need to work quickly to minimize stress to the bird (not to mention to yourself). All the instruments you'll need come in a caponizing kit (at least one source is listed in the appendix). The kit should include forceps or testicle removers, a sharp knife, a rib spreader, and a sharp pointed hook. Make up a solution of 10 drops of chlorine bleach in a quart of water to rinse your tools between birds.

A helper comes in handy to hold the birds while you work. If you work alone, tie birds down so they can't flap. The traditional way to hogtie a bird for caponizing is to use two cords about 2-feet (60 cm) long, each fastened at one end to a 1-pound (0.5 kg) weight (window-sash weights or something similar). Using a half hitch, tie the cockerel's legs with one cord. Again with a half hitch, use the other cord to tie the wings together near the shoulder joint. Keep the bird stretched out by hanging the weights in opposite directions over the edges of your work table.

Begin by moistening and removing feathers from a small area over the last two ribs on one side, just in front of the thigh. With one hand, slide the skin down toward the thigh and hold it so you won't cut into the muscle and cause bleeding. Make a small cut in the skin, parallel to and between the last two ribs. Lengthen the opening in both directions until it is about 1-inch (2.5 cm) long.

Insert the rib spreader into the cut and use it to spread the ribs apart. Through the opening you'll see the intestines beneath a thin membrane. Use your hook to tear the membrane and expose the near testicle just below the membrane. The testicle is that white or yellowish (or maybe darker) lump, about the size and shape of a navy bean, close to the backbone.

If the intestine is empty, as it should be, push it aside to find the far testicle. It's in the same position as the first one, but on the opposite side of the backbone. If you can easily find the lower testicle, remove it first. If you can't readily find it, remove the upper testicle, turn the bird over, and remove the second testicle through a second incision.

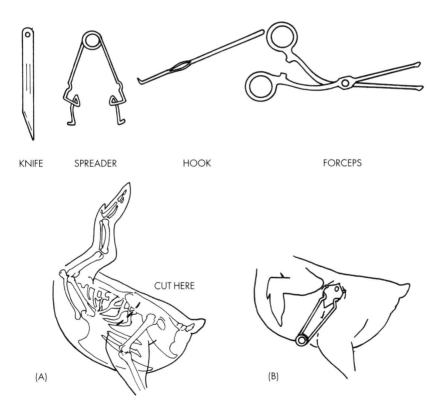

KNIFE   SPREADER   HOOK   FORCEPS

CUT HERE

(A)   (B)

A caponizing kit includes a knife, rib spreader, sharp hook, and forceps. (A) Use the knife to make an incision between the last two ribs. (B) Spread the ribs with the rib spreader, tear the membrane with the hook, and lift out the testicle with the forceps.

To remove a testicle, grasp it with the forceps and slowly twist, breaking the organ free of the attached cord. Watch out for two things: that you grasp and remove the *entire* testicle, and that you do not accidentally grasp the artery just behind the testicle. A bird with a ruptured artery will bleed to death. Take heart: the chance that you'll rupture an artery is minimized if you withheld feed and water prior to the surgery.

You needn't stitch the incision shut. When you remove the rib spreader, the skin will come back together. Pour a little hydrogen peroxide over the wound and, if it makes you feel better, coat it with a dab of Neosporin antibiotic ointment. The wound will completely heal within 10 days. Since caponizing causes stress, add a water-soluble vitamin supplement to the birds' drinking water for a few days.

Caponizing Problems

Caponizing cockerels without making a mistake requires practice. If you chance to kill a bird, don't feel bad about it — even the most experienced caponizer occasionally loses one. Despite it's small size, the bird may still be dressed and cooked for dinner.

If, in your first few tries, you leave a portion of a testicle behind, you'll end up with what's known as a "slip." The piece of testicle left behind will continue to grow, making the bird neither capon nor cockerel. Slips don't fatten much faster than a cockerel and are just as restless and aggressive. They may chase mild-mannered capons away from feed and water, reducing their growth rate. You'll know you have a slip if its head and wattles redden instead of remaining pale. Raise and butcher a slip as if it were an intact cockerel.

Within the first ten days after caponizing, air may accumulate beneath a bird's skin, causing the skin to balloon out (a condition colloquially called "windpuff"). If that happens, prick the skin with a needle or the point of a sharp knife and press out the air.

Because of their weight, capons are more susceptible to breast blisters than other birds. Since capons are so heavy, merely resting their keels on a hard surface, such as dirt or packed litter, can cause breast blisters. To minimize the problem, house capons on loose, dry, fluffy litter and don't let them perch.

Leg injury, another common capon problem, occurs when birds are fed for rapid growth so that, quite literally, they get too heavy for their little legs to carry them. Range-fed capons are less likely to have weak legs than confined capons. You can tell capons are developing weak legs if they spend a lot of time sitting around.

Feed your capons a 17 percent grower ration followed by a finisher for the last two weeks. A capon should weigh — give or take a little to allow for the more rapid growth of meat strains compared to noncommercial strains — 3½ to 4 pounds (1.6–1.8 kg) by 8 weeks of age, 7 to 8 pounds (3.2–3.6 kg) by 20 weeks, and 9 to 10 pounds (4–4.5 kg) by 25 weeks.

A well-fleshed capon will eat 24 to 38 pounds (11–17 kg) of feed during its lifetime, so you can see why it doesn't pay to feed capons beyond their peak weight at 25 weeks for nonhybrids, 20 weeks for hybrids. If for some reason you should choose to feed them longer, rest assured they will remain tender.

Meat Bird Quality

A bird's readiness for butchering and the quality of its meat depend on:

- ◆ Freedom from defects — no crooked breast bones, crooked or hunched backs, deformed legs and wings, bruises, cuts, tears, breast blisters, or calluses.
- ◆ Feathering — mature feathers only, no stubby broken feathers or pin-feathers.
- ◆ Fleshing — degree of meatiness on the breast, legs, and thighs.
- ◆ Finish — layer of fat beneath the skin. (Spread breast feathers to examine the skin. A creamy or yellow color indicates good finish; a reddish or bluish color indicates too little fat.)
- ◆ Conformation, or body shape — the ideal is blocky and rectangular, not narrow and triangular:

Breast Fleshing

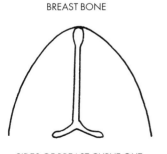

BREAST BONE

SIDES OF BREAST CURVE OUT

Ideal

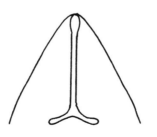

SIDES OF BREAST ARE FLAT

Acceptable

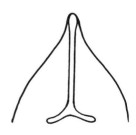

PROMINENT BREAST BONE

SIDES OF BREAST CURVE IN

Poor

Economic Efficiency

The poultry industry uses several different methods to determine the economic efficiency of producing meat birds. Since the cost of feed accounts for at least 55 percent of the cost of meat-bird production, most economic indicators factor in the amount of feed used.

Feed conversion ratio is the total amount of feed in pounds (or kilograms) eaten by the birds divided by the flock's total live weight in pounds (or kilograms). Commercial broilers raised under efficient methods get by on as little as 1.85 pounds of feed per pound of weight gained to 4 pounds live weight. At home, don't expect a conversion ratio from a meat strain much better than 2 pounds of feed per pound of live weight, or about 3 pounds of feed per pound of dressed weight. If your rate is considerably higher, take stock of your management methods.

Feed cost per pound (or kilogram) is the feed conversion ratio multiplied by the average cost of feed per pound (kilogram). Determine the average cost of feed per pound by dividing the total pounds of feed used into your total feed cost. Alternatively, determine the average cost of each 50-pound bag of feed you used and find your feed cost per pound on the accompanying chart. The lower this number is, the better you're doing.

Performance efficiency factor is the average live weight divided by the feed conversion ratio, multiplied by 100. In industry, this index hovers around 200. The higher you get above 200, the better you're doing.

Livability is the total number of birds butchered or sold divided by the total number started. To convert that number to a percentage, multiply by 100. Good livability is 95 percent or better. If your livability is above 90 percent, you're doing as well as many commercial growers.

Average live weight is the total live weight divided by the total number of birds. The industry average for hybrids efficiently raised to 8 weeks of age is 4 to 5 pounds (1.8–2.25 kg). If your birds aren't even close at the end of 8 weeks (13 weeks for nonhybrids), look for ways to improve your management methods. One way to improve your average is to raise cockerels instead of straight-run chicks. As a general rule, finished cockerels will weigh 1 pound more than pullets at the same age and on the same amount of feed.

Average weight per bird is the total dressed weight of all birds divided by the total number of birds. This index factors in weight lost both to excess fat and to uncontrollable inedible portions such as intestines, feathers, heads, feet, and blood. A good average for the edible portion is approximately 75 percent of a bird's live weight.

Feed Cost per Pound of Live Weight*

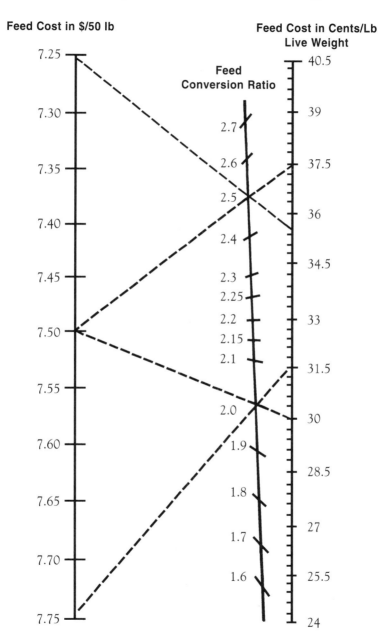

Feed Cost in $/50 lb

Feed Cost in Cents/Lb Live Weight

Feed Conversion Ratio

*Place a ruler across the price per 50 lb bag of feed and feed conversion ratio to find the feed cost per pound of live weight.

Adapted from: *Cobb Broiler Manual*, Cobb, Inc.

CHAPTER 5

BUTCHERING DAY
(AND GOOD EATING TO FOLLOW)

THE WORST PART about killing chickens is the reflex reaction that causes them to flap and twitch for a few moments after they're dead. During these death throes, blood gets spattered around. If butchering is going to upset you, flapping and bleeding are the most likely aspects to do it. That goes double for children, who may not understand that even though a bird is still moving, it's dead and isn't suffering. Nearly every family has at least one member (or neighbor) who can handle killing chickens.

Butchering

When your birds are plump enough to butcher, catch a few and examine their pinfeathers. Chickens with emerging pinfeathers won't look clean no matter how carefully you pick them. Catch birds at night, when they're least active, to minimize struggling. Struggling depletes energy a bird needs to relax its muscles and keep the meat tender.

Confine birds overnight with plenty of water but without feed. Put them in a wire- or slat-floored coop where they can't eat feathers or litter. The goal of cooping them without feed is to allow their intestines to empty, minimizing contamination during butchering. They need water, however, to prevent dehydration, which turns the skin dark, dry, and scaly, and causes meat to get tough and stringy.

Killing

Set up a place to kill chickens away from the rest of the flock, where remaining birds won't get upset and where you can easily clean up afterwards to avoid attracting flies, wasps, and four-legged predators. Have on hand a can or bucket lined with a sturdy plastic bag, so you won't be tempted to toss discarded parts on the ground.

Bleeding is an important part of the killing process to ensure that no trace of blood remains in the meat. It must be done while blood still flows freely. Be ready to deal with blood, either by working where it can easily be flushed away or by having a collection container on hand. If you plan to compost the blood, a little water in the collection container will keep it from coagulating. Have additional water handy to dilute spills.

Of several possible methods for killing chickens, the one most commonly used in backyards is the least suitable.

Ax. Most people's image of killing a chicken is to lay its neck on a stump and chop off its head. Using an ax is not the best idea, however. An ax severs the bird's spinal cord and hinders bleeding. It also severs the jugular vein at the same time as the windpipe, letting blood into the lungs and contaminating the meat.

Hand. A more suitable time-honored method is to dislocate the bird's neck, which kills it instantly. Hang the bird upside down by holding both its shanks in one hand. With the other hand, grasp the head with your thumb behind the comb and your little finger beneath the beak. Tilt the head back and pull steadily until the head snaps free of the neck. Continue holding the head until struggling stops.

If you have trouble snapping a chicken's neck by hand, as I do, use your feet: Grasp the bird by the shanks and lay its neck on the ground. Place a broom or rake handle across the neck. With one foot on each side of the neck, stand on the handle and firmly pull the bird upward.

After the neck has been snapped, hang the bird by its legs with a piece of twine, cut the neck on both sides, and let it bleed.

Knife. If you have a lot of birds to kill at one time, using a knife lets you work fast. Suspend each bird from twine tied around its legs. With one hand, grasp the beak in front of the eyes and pull downward to hold the bird steady. With a sharp knife in the other hand, make a 2-inch (5 cm) cut just behind the jaw and into the base of the skull on both sides of the neck, severing the jugular vein.

Debraining helps loosen feathers for hand picking. After cutting the vein for bleeding, insert your knife into the mouth, its sharp edge toward the groove at the roof. Push the knife toward the back of the skull and give it a one-quarter twist. The trick is to avoid sticking the *front* of the brain, which causes feathers to tighten instead of loosen. You can tell your knife hits home when reflex causes the bird to shudder and utter a characteristic squawk.

Gun. At our place, we use a .22 handgun because it's fast and clean. This is not an option to consider unless you're familiar with guns and you live in a rural area where shooting is legal and can safely be done. Hang the bird by its legs and shoot a round into the back of the head. Use a knife to slit the jugular vein for bleeding.

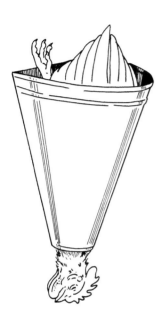

To keep blood from spattering, you have at least three choices:

◆ Keep a tight hold on the bird until flapping and bleeding stops.

◆ Hang a 1-pound (0.5 kg) weight (a window-sash weight or something similar) from a sharp hook attached to its lower beak.

◆ Confine the bird in a killing cone — a funnel-shaped device into which a chicken is inserted, head first, to keep it still during killing. Cones come in various sizes; choose the size that fits your birds snugly. (Cones and other butchering supplies are available from some of the sources listed in the appendix.)

The killing cone is also called a holding funnel.

Picking

Once the bird is dead, you have to decide whether to remove the head before or after picking. The "right" way is to leave the head on until the feathers are removed, which keeps the neck clean for making soup, broth, or gravy. Since I don't like to have a bird stare accusingly at me while I relieve it of its feathers, I behead it first.

The process of removing feathers is called "picking" in the industry and "plucking" in most other circles. For sanitary reasons, it makes good sense to kill and pluck in one area, then rinse and pack in another. If you have a helper, the whole operation will go quickly if one does the picking and the other does the packing.

Not being one to follow the conventional wisdom that says picking is less messy if you do it outdoors, I prefer to work indoors where it's cool and I don't have to fight off flies and yellow-jackets. I kill four birds at once and bring them inside for picking. Four is as many as I can do before the last one starts getting stiff and hard to manage. When I'm done, I go back and get four more. After the whole batch is picked, I clean up before packing the birds away.

Depending on how often you butcher and how many birds you have, you may pick in one of four ways:

Hand picking involves pulling the feathers out by hand, which can be done by:

- the dry-pick method
- the scald and pick method (discussed in the next section).

Dry picking is suitable when you have only a few birds. It's easy when the birds are freshly killed and still warm, especially if they've been debrained. But it's not so easy after birds start to cool. The best plan is to strip the feathers away quickly while the birds are still hanging for bleeding. If you're allergic to feathers or dander, wear a dust mask during dry picking.

To remove pinfeathers — the immature feathers remaining after you have removed the main feathers — you can use:

- a pinfeather scraper — a 3- to 4-inch (7.5–10 cm) knife with a round end
- your fingers and/or a dull-bladed paring knife
- a pinfeather puller — basically a pair of tweezers
- wax

Wax picking follows rough hand picking as a fast way to get birds clean. It involves dunking the whole bird in hot wax, then dunking it in cold water to harden the wax. Peel away the wax, and feathers and pinfeathers alike come right off. When you're all done, let the remaining wax cool in the pot. Debris will settle to the bottom, and you can save the clean wax on top for reuse. Picking wax is sold by suppliers of butchering equipment. Wax picking is used less often for chickens than for ducks and geese, which are much harder to pluck.

Machine picking is best suited for large-scale operations, since picking machines don't come cheap. If local laws allow, you might defray the cost by doing custom picking on the side or by letting others use your machine on a rental basis. A picking machine has rotating rubber fingers that rapidly flail feathers off, which sounds nifty until you learn that a University of Arkansas study found mechanically picked chickens to be 2.5 times tougher than hand-plucked birds.

Custom picking is no longer common in some areas, but if you find the idea of killing your own chickens too messy or distasteful, you might look for a custom butcher or picker. As an alternative, seek out someone who cleans game birds for hunters. Make sure the person knows how to deal with live birds, though: One fellow I know who regularly picked dead birds for hunters happily killed a bunch of chickens for a friend but didn't have the sense to bleed them.

Scald and Pick

Birds that have started to cool are easier to hand pick if you loosen the feathers by dipping each bird in scalding water. A squirt of dishwashing detergent in the water helps penetrate dense feathers. Hold the bird by the shanks and completely immerse it, head first, until all its feathered parts are under water. Move the bird up and down and back and forth to help the water evenly penetrate the feathers.

The kettle must be large enough so when you dunk and slosh, the rising water won't overflow. A standard-size enamel canning pot two-thirds full is just right unless you're doing capons, in which case you'll need an outsized pot. If you plan to do a lot of birds, you might invest in a thermostatically controlled dipping vat.

Loosening the feathers of older, tougher birds requires a higher temperature and longer scalding time than is needed for young tender birds, as indicated in the accompanying chart. Unfortunately, scalding at temperatures higher than 135°F (57°C) stiffens muscle tissue, making tough meat even tougher.

Test the scald by pulling a few tail and large wing feathers. If they don't slip right out, quickly dip the bird again. Hang the bird by its legs or lay it on a table for picking. Working as rapidly as possible, pull handfuls of feathers, rubbing with the base of your thumb as you work. Feathers should strip off easily. If they don't, you didn't dunk long enough, the water didn't penetrate evenly, or the water wasn't hot enough. Torn skin and patches of feathers coming off with attached skin are signs that you dipped the bird too long or the water was too hot.

Pick the feathers in the sequence that works best for you. I like to start with the large wing and tail feathers, then pick both wings clean, then work down the breast and legs, and finally move down the back to the tail.

The world picking record is 4.4 seconds per bird. The United States Department of Agriculture (USDA) claims that an experienced picker shouldn't take more than 5 minutes for each bird. Call me slow, but after 20 years of experience, my hand-picking time is still about 15 minutes per bird.

Once the main feathers are off, squeeze out pinfeathers by holding the bird under a running faucet and scraping the skin with a pinning knife or a dull-bladed paring knife. Stubborn pins may need to be tweezed out or pulled be-tween a paring knife and your thumb. When the skin is clean, rinse off loose feathers, pat the bird dry, and set it aside while you pick the next.

Scalding Temperatures

Class	Water Temperature		Time
Broiler/fryer	128–130°F	53–54°C	30 seconds
Roaster	128–130°F	53–54°C	30 seconds
Capon	128–130°F	53–54°C	30 seconds
Stewing Hen (and other old birds)	155–160°F	68–71°C	30–60 seconds

Singeing

If you raise utility strains and you serve chicken with the skin on, you'll need to singe away the hairlike feathers that pop up after birds have been plucked and dried. Hairs singe off readily when passed over an open flame.

I used a gas burner until I moved to a house with an electric stove. Then I tried a candle, but it made the skin sooty. Eventually I settled on a propane torch, held by my husband while I moved each chicken back and forth. We didn't learn until later that singeing torches, similar to a handyman's propane torch, are used commercially.

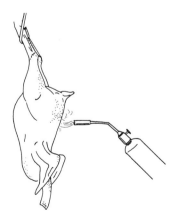

Use a gas flame or propane torch to singe off hairlike feathers.

These days we don't eat the skin (cholesterol, you know), so I no longer worry about hairs. If I'm going to freeze the meat, I leave the skin on as protection against freezer burn. Since I peel the skin off before cooking the meat, no one need know my birds are hairy. If you raise a Cornish cross, you don't have to worry about singeing, anyway, since hybrids have had their hairs genetically removed.

Eviscerating

Some people clean birds right after they're plucked. Others chill birds first, finding it easier to remove the internal organs from a cooled bird. Try both ways and see which one you prefer.

If you butcher regularly, a boning knife or a pair of boning shears comes in handy, or you can use clean sharp pruning shears. Use the knife or shears to remove the feet at the hock joint. If you're cleaning an older bird — roaster, capon, or stewing hen — avoid cutting the tough tendons so you can pull them from the drumsticks: Cut the skin along the back of the shank to expose the tendon, insert a hook, twist the hook to get a good hold on the tendon, and pull. A tie-wire twister, used to twist concrete tie wires, makes a dandy tendon remover.

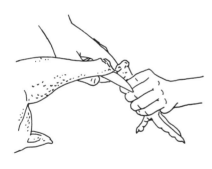

(1) To remove feet, cut through the leg just below the hock joint.

To remove the feet, cut through the leg just below the hock joint. Unless you've already cut the back to remove the tendons, work from the front toward the back to leave a little flap of skin that keeps the meat from shriveling away from the bone during cooking.

Turn the bird over to cut off the oil gland on the tail. One inch (2.5 cm) above the gland's nipple make a cut deep enough to go to the tailbone. Since the gland goes quite deep, cut with a scooping motion as you move the knife along the bone to the tip of the tail. Be sure to get the whole gland out so it won't taint the surrounding meat.

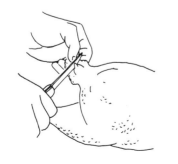

(2) Remove the oil gland with a deep wedge-shaped cut.

If you haven't already done so, remove the head next. Use your knife or shears and, with a twisting motion, cut between the head and the first neck vertebra.

If you plan to stuff the bird, leave a flap of neck skin to hold the stuffing in and keep it from drying out. Insert the knife through the skin at the back near the shoulders. Slit the skin by guiding the knife up the back of the neck. Pull the skin away from the neck down to the crop, and cut below the crop to remove the crop and the windpipe. (The crop will have no contents to spill if the bird hasn't had access to feed for at least 12 hours.)

Use your kitchen shears to cut the neck all the way around, as close as possible to the shoulders. Grasp the neck and twist it off. If you can't get a good grip, wrap the neck in a dry paper towel. Fold the flap of skin back and use the wing tips to pin it down.

If you're going to cut up the bird for cooking, cut off the skin at the base of the neck, remove the crop and windpipe, and cut and twist off the neck without leaving a skin flap. Although the neck doesn't have much meat, it can be used to flavor broth.

(3) Slit the skin along the back of the neck.

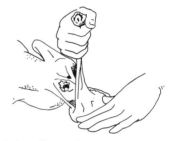

(4) Pull out the crop and the windpipe.

Opening the Abdomen

The way you open the abdomen to remove the internal organs will depend on the bird's conformation and on whether or not you plan to roast it whole.

A *bar cut* offers a natural trussing method for whole birds by leaving a horizontal strip of skin across the abdomen that you can tuck the

(5) Cut through the neck at the base and twist it off.

legs into, to keep the breast meat from drying out. A bar cut doesn't work for birds with large deposits of fat, which loosen the abdominal skin too much to hold the legs, and it doesn't work for birds with legs that are too short to tuck in.

Use a sharp, thin knife to separate the vent from the tail, halfway around. Do so by inserting the knife between the vent and tail and working upward, one-quarter around until you reach the pointed bone at the side of the vent (pinbone). Then work upward in the other direction toward the other pinbone, taking care not to cut into the intestine.

Insert your finger into the opening as a guide and use your shears to continue cutting the skin away from the vent, all the way around. Grasp the vent and pull it out a little so it can't drop into the cavity and release fecal matter inside the bird.

About halfway between the vent opening and the breast bone, pinch the skin and insert your knife to make a horizontal cut about 3 inches (7.5 cm) across. The strip of skin between this cut and the vent opening is the skin bar that holds the legs in place during roasting.

A *midline cut* is a vertical opening running from the vent straight to the breastbone. This is the cut used for broiler/fryers and small-sized roasters. Roasters opened this way must be trussed with string or skewers to hold in stuffing or to keep the breast meat of an unstuffed bird from drying out.

Stretch the abdomen skin with one hand and use a knife to cut through the skin with the other hand, inserting the knife at the keel. Slowly work the knife toward the vent, taking care not to cut so deep that you break into the intestine. When you reach the vent, insert your index finger into the opening and lift up on the intestine while you continue cutting around the vent, beneath your finger.

Opening the Abdomen

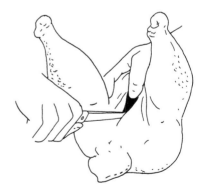

Bar Cut: If the bird is to be roasted whole, slit horizontally 2 inches (5 cm) from the vent opening.

Midline Cut: If the bird is to be cut into pieces, slit vertically between the vent and the keel.

Terminology

Eviscerating a chicken is not as complicated as you might infer from the terminology used in government publications and other manuals. Knowing what all those strange words mean makes the whole thing look as simple as it is.

Viscera — internal organs
Entrails — another word for viscera
Giblets — edible organs (liver, heart, gizzard)
Offal — inedible organs (crop, intestines, etc.)
Eviscerating — removing the organs
Drawing — another word for eviscerating

Drawing

Once the abdomen is properly opened, the next step is to remove the internal organs or viscera. Insert a hand into the opening and work it around the inside wall, as far as you can reach on both sides, to break the attaching membranes. When you come to the gizzard (it feels hard in comparison to other organs), cup your hand around the bundle of organs and pull gently. Although the idea is to bring out all the organs at once, if you miss any you can always reach back in for them. When you pull them out, the organs will still be attached to the bird. For the moment leave them that way.

Evisceration

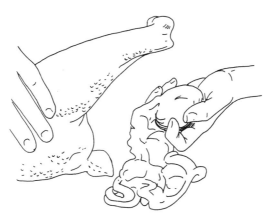

Remove the internal organs in one bundle.

If you're cleaning a cockerel, be sure to remove the testicles — two soft, white oval-shaped organs along the backbone. If you're cleaning a pullet or hen, remove the mass of eggs located in the same area.

In both cases, remove the lungs — pink spongy organs pressing against the upper back (some people enjoy the lungs in soup, provided they are not contaminated with blood during killing). You can easily remove the lungs by inserting an index finger and lifting them with a scraping motion.

Giblets

In the bundle of organs are three things you may want to save: the liver, the heart, and the gizzard. The liver is delicious, provided you don't contaminate it by breaking the gall bladder — the small green sac nestled into a fold. You'll know the bladder has broken if the liver becomes stained with green bile. If that happens, throw the liver out, as it will taste bitter.

Some people remove the gall bladder by pinching it between their thumb and index finger. I've never been able to do that without squeezing out bile. Instead, I carefully insert a sharp knife tip under the connective tissue and slice upward, pressing the end of the bladder between my thumb and the blade to keep it from spilling its contents.

These days, a lot of people don't take the time to save the heart and gizzard, but if you clean a lot of chickens, gizzards and hearts can add up to a heap of good, slightly chewy meat. Mince it to add to gravy or stuffing, or grind it up for pizza, spaghetti sauce, tacos, or chili. Even when I don't save hearts and gizzards to serve at the table, I cook them up for my appreciative cat.

Remove the membrane around the heart, trim off the top, slit the heart halfway open, and rinse it. Cut away the gizzard where it attaches to the stomach and intestines. Cut into the large end, aiming toward the center until you come to the tough lining. In an old bird you can separate the gizzard from the

Cleaning the Gizzard

Cut into the large end of the gizzard until you reach the tough lining.

Peel away the lining and discard it, along with its gritty contents.

Mince or grind the remaining meat; cook it for use in recipes or pet foods.

lining without cutting into the latter. If the lining is tender enough to tear, cut into it and peel it off with your fingers. The lining will be full of grit that needs to be rinsed away.

Some people wrap the giblets from each bird in a little plastic bag and place it in the body cavity. I prefer to separate giblets for packaging. I grind the hearts and gizzards and package them together. I separate the livers to cook with bacon and/or onions. Liver doesn't keep well, so I like to serve it fresh. If you have too many livers at once, pack them in plastic freezer containers large enough to hold one meal each. Cover the meat with a piece of plastic wrap or waxed butcher paper before putting on the lid, and freeze.

Now you can break the attachments connecting the remaining organs. Drop this refuse into a large bucket or cardboard box lined with a sturdy plastic bag. In most areas, regulations prohibit disposal of offal in dumpsters or landfills. If you have a place to bury it away from water sources, bury it at least 3 feet (90 cm) deep. Otherwise compost it.

Components of a Broiler/Fryer

Component	Live Weight		Percentage
Total	4.1 lbs.	1.86 kg	100.00
Blood	0.170	0.0772	4.14
Feathers	0.332	0.1510	8.10
Head	0.097	0.0441	2.37
Feet	0.160	0.0726	3.90
Offal	0.392	0.1780	9.56
Giblets	0.401	0.1820	9.80
Cleaned bird	2.550	1.1580	62.20

Source: *Cobb Broiler Manual*, Cobb, Inc.

Cooling

As soon as the birds are cleaned, thoroughly rinse them in running water and cool them quickly to remove remaining body heat and check bacterial growth. Put them in a clean container full of cool tap water. Either let the water slowly overflow the container, or change the water several times until the birds are cool.

Proper cooling lets you avoid a major source of contamination found in store-bought chickens. Commercial birds are cooled in vats of water laden with rinsed-away fecal matter spilled from intestines that have been torn open by eviscerating machines. Meat soaks up the foul water and bacteria.

Unfortunately, you can't avoid bacteria altogether. The little critters are everywhere. To minimize bacterial growth, get the meat cooled and refrigerated as quickly as possible. When the birds are cool, remove them from the water and either prop them on paper towels or hang them by a wing to drain for about 20 minutes before wrapping them for aging.

Approximate Yields

Type Bird	Yield	Waste
Roaster or capon	76%	24%
Broiler/fryer	75%	25%
Stewing hen, heavy breed	70%	30%
Stewing hen, Leghorn-type	68%	32%

Aging

Freshly killed chicken, like other meat, must be aged. Aging allows muscle tissue to relax after it has stiffened as a result of muscle protein coagulation soon after death. An aged chicken is tastier and more tender than a chicken cooked or frozen soon after being killed.

Wrap chickens loosely and age them in the refrigerator for a day or two, taking care to space them so cool air can circulate around them. If you plan to cook the chicken fresh, you can keep it refrigerated for up to five days. If you plan to freeze or can it, do so no later than the third day after it was dressed.

If you butcher more birds than your refrigerator can properly chill, place them in ice water until they reach 40°F (4°C). Since chilling time varies from around 6 hours for a fryer to 10 hours for a big roaster, you'll need plenty of ice.

Cutting Up

Fryers are usually cut into quarters or smaller pieces for frying. Broilers may be left whole for open-pit roasting or cut into halves or quarters for barbecuing. Roasters and capons are generally left whole. Old hens, being too tough to roast, are cut up to be stewed or fricasseed.

You may occasionally run across directions for eviscerating and cutting up a bird in one operation, but cutting up a chicken soon after it has been killed causes its muscles to bunch up, making them dense and tough. A chicken that is eviscerated and aged before being cut up will be more tender. Besides, chilled meat is easier to handle.

How to Cut Up a Chicken

To cut up a chicken, use a sharp, heavy knife and follow these steps:

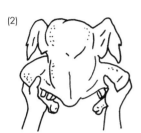

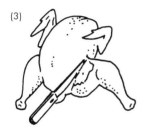

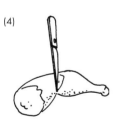

1. Cut skin between thighs and body.
2. Grasp a leg in each hand, lift the bird, and bend the legs back until the hip joints pop free.
3. Cut leg away by slicing from the back to the front at the hip, as close as possible to the backbones.
4. If you wish, separate the thigh from the drumstick by cutting through the joint between them. You can find the joint by flexing the leg and thigh to locate the bending point.
5. On the same side, remove the wing by cutting along the joint inside the "wingpit," over the joint and down around it. Turn the bird over and remove the other leg and wing. To create mini-drumsticks, separate the upper, meatier portion of each wing from the lower two bony sections.
6. To divide the body, stand the bird on its neck and cut from the tail toward the neck, along the end of the ribs on one side. Cut along the other side to free the back. Bend the back until it snaps in half, and cut along the line of least resistance to separate the ribs from the lower back.
7. Place breast on the cutting board, skin side down, and cut through white cartilage at "v" of the neck.
8. Grasp the breast firmly in both hands and bend each side back, pushing with your fingers to snap the breastbone. Cut the breast in half lengthwise alongside the bone. For boned breasts, place the breast skin side up on a cutting board. Insert the knife along one side of the keel and cut the meat away from the bone. Repeat for the other side.

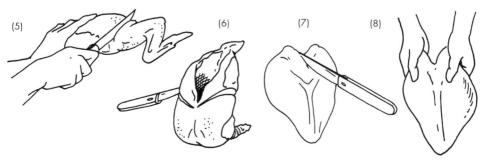

If you wish, sort the meatier pieces (breasts, thighs, and drumsticks) from the bonier pieces. Package the meaty pieces for meals and strip the bony parts for canning or soup.

Usable Parts of a Broiler/Fryer

Part	Approx. Percentage, Dressed Weight
Legs & thighs	37%
Breast	34%
Wings	13.5%
Back & neck	13%
Fat	2.5%

Equipment Sanitation

Cleaning chickens is a messy affair and is best done somewhere other than the kitchen. No matter where you do it, after handling raw meat, thoroughly clean countertops, sinks, and any utensils you use. To control bacteria that cause spoilage and food poisoning, any item that touches meat must be as clean as possible.

◆ Wash hands, knife, and other utensils with hot soapy water. Rinse utensils in boiling water.
◆ Scrape bits of meat from counters and cutting boards, then wash these surfaces with hot soapy water and rinse with boiling water.
◆ Sanitize counters and cutting boards with a solution of 2 teaspoons chlorine bleach in 1 quart (10 ml per liter) of warm water. Leave the solution in contact with surfaces for 20 minutes before rinsing it away with boiling water.

Cooking the Bird

If you're well known for your lip-smacking chicken, you might try your hand at one of the numerous annual contests held throughout the country. Prizes run highest in the National Broiler Council's National Chicken Cooking Contest (listed in the appendix).

The infinite number of recipe variations can be distilled into six basic ways to cook chicken: microwave, fry, broil, braise, stew, roast.

To microwave, dip parts in melted butter, roll in seasoned crumbs, cover parts loosely with wax paper or plastic wrap, arrange meatier parts toward outside of dish and boney parts toward the center, rearrange or turn parts halfway through microwaving time, let microwaved chicken stand 5 minutes, add salt during standing time (not before microwaving), check for doneness after standing time. The accompanying chart shows settings and times for one or two pieces; consult your oven's manual if you wish to cook more pieces at one time.

Microwaving Timetable

Parts	Power	Minutes
1 breast half	high	4–6
2 breast halves	high	8–10
1 thigh	medium-high	4–5
2 thighs	medium-high	6–8
1 leg with thigh	medium-high	7–9
2 legs with thighs	medium-high	9–13

Source: National Broiler Council.

To fry, heat just enough oil to cover the bottom of a frying pan. Quickly brown pieces (plain or coated with batter, crumbs, or spices) on all sides, then lower the heat and cook 30 to 45 minutes. For a crisp coating, cover the pan only with a splatter guard; for a softer coating, cover with a lid until the last 10 minutes. Drain pieces on brown paper or paper towels.

To deep fry, coat pieces with flour, cracker crumbs, wheat germ, or a thin batter. Heat a few pieces at a time in oil at 350°F (177°C). Drain well on brown paper or paper towels.

To oven fry, coat pieces with flour, cracker crumbs, wheat germ, sesame seeds, or batter. Bake in a shallow pan, with the oven set at 350°F (177°C), for 30 to 45 minutes.

To broil, place halves or pieces on a rack in an oven set on broil, 4 to 6 inches (10–15 cm) from the heat, or on an outdoor grill kept at low to moderate heat. Allow 20 to 30 minutes per side, turning once. Leave the meat plain or baste it toward the end with your favorite sauce.

To braise, simmer pieces slowly in a covered pan with a small amount of broth, fruit juice, or other liquid until meat is tender.

To *stew*, cover stewing-hen pieces with water and simmer, covered, for 2 to 3 hours (or cook under pressure according to directions that come with your pressure cooker). For a younger bird, simmer only 45 minutes to 1 hour. Add vegetables, fruit, and/or spices for additional flavor.

To *roast*, put the whole bird, breast side up, on a rack in a roasting pan. Roast a stuffed bird at 325°F (163°C), an unstuffed bird at 400°F (204°C), uncovered, until the thigh is tender.

Roasting Timetable

Kind	Weight in Pounds	Time at 325°F (166°C)	Time at 400°F (204°C)
Pieces	¼–¾	1–2½ hours	
Rock Cornish hen, stuffed	1–2	1–2	
Broiler/fryer, unstuffed	1½–2½		¾–1½
Broiler/fryer, stuffed	2¼–3¼	1¾–2½	
Roaster, unstuffed	2½–4½		1½–2¾
Roaster, stuffed	3¼–4½	2–3	
Capon, stuffed	4–8	3–5	

Chicken is fully cooked if:

1. It feels tender when stabbed with a fork.
2. The drumstick wiggles freely at the joint when lifted or twisted.
3. The thickest part of the drumstick feels soft to your fingers.
4. Juices run yellow, not pink, when you pierce or cut into the drumstick.
5. A meat thermometer inserted into the thickest part of the thigh, not touching the bone, registers 190°F (88°C).

Roasted chicken is always a favorite.

Trussing

Trussing birds to be roasted whole gives them a nice, compact appearance. You can truss a bird by one of three methods:

Natural truss is the neatest method, but is possible only for birds opened with a bar cut. Fold the neck flap over the back and bring the wing tips across to hold the flap in place, then tuck the hock joints under the skin bar. To seal the vent opening, slit the tail through to the vertebra at the back, bend the tail forward, and push it through the vent opening.

String truss the bird by running a clean string from one shoulder to the other over the wings, then across the back, around the drumsticks, and across the back again. Tie the string tightly around the tail. To hold in stuffing and keep it from drying out, you may have to stitch the opening shut with a needle and string.

Air truss involves shaping the bird prior to roasting. Place the bird in a plastic bag with its legs pressed close to the body. Use a vacuum to draw out the air, then seal the bag so the bird retains its shape.

If you don't stuff the bird, take advantage of an old Chinese trick that decreases cooking time by speeding up heating. Before trussing the bird, place a couple of metal spoons inside. The spoons will heat up, roasting the bird from the inside as well as the outside. Gourmet kitchen shops carry steel racks that improve on this method by holding the bird in a vertical position, letting fat drain away while conducting heat to the inside.

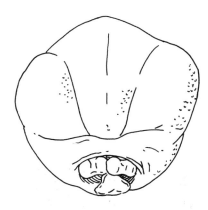

The neatest trussing method is the natural truss, for which the bird must be opened with a bar cut.

Nutritional Value

Chicken is a short-fiber meat that is easily digested by children, older people, and those with digestive problems. The nutritional value varies with the cut and the method by which it is cooked. Compared to roasting or stewing, frying increases calories and fat. Dark and light meat are similar in vitamin and mineral content, but dark meat is higher in fat and calories than white meat.

Fat varies more than any other nutritional component. Abdominal fat is the most variable fat deposit and therefore the main measure of fatness. Chickens evolved with the ability to deposit large amounts of abdominal fat to use as reserve energy during times when forage is scarce. Excess abdominal fat in a domestic flock represents wasted feed and wasted money. Older birds generally contain more fat than younger ones, and females have more fat than males. As the percentage of fat increases, percentages of protein, minerals, and vitamins decrease.

Nutritional Composition

100g Serving	Calories	Protein	Saturated Fat	Cholesterol
Light meat				
fried, batter-dipped	260	24.8 g	3.5 g	85 g
fried, flour-coated	222	31.8 g	2.5 g	89 g
stewed with skin	184	27.4 g	2.0 g	75 g
roasted with skin	197	29.8 g	2.2 g	89 g
roasted without skin	173	30.9 g	1.3 g	85 g
Dark meat				
fried, batter-dipped	273	21.8 g	4.3 g	90 g
fried, flour-coated	254	26.8 g	3.9 g	94 g
stewed with skin	233	23.6 g	3.6 g	82 g
roasted with skin	253	25.7 g	4.0 g	91 g
roasted without skin	205	27.4 g	2.7 g	93 g

The difference between similar cooking methods with or without the skin is negligible if you don't eat the skin.

Chicken Meat Safety

To keep your homegrown chicken safe:

◆ Thoroughly rinse meat in clear water after butchering.
◆ Heat meat to no less than 185°F (85°C).
◆ Cook casseroles and other combination foods to 160°F (71°C).
◆ Barbecue chicken until no red meat appears at the joints.
◆ Never return barbecue and other cooked meat to the plate on which it was carried while raw.
◆ Hold cooked chicken above 140°F (60°C) for no longer than two hours before serving it.

◆ Add the stuffing to a whole chicken just before you pop the bird into the oven.

◆ Separate meat from stuffing or gravy before storing leftovers in the refrigerator.

◆ Reheat leftovers to at least 165°F (74°C).

◆ Before cooking frozen meat, thaw it in the refrigerator or in a plastic bag under cold running water, never on the counter top.

Chicken Meat Storage Safety

	Degrees Centigrade	Degrees Fahrenheit	
	100	212	Water boils
		180	Whole chickens and chicken parts cooked done
		170	Chicken breast cooked done
	70	158	Maximum incubation temperature for thermophilic bacteria
	60	140	Minimum safe holding temperature for hot chicken
	40	104	Minimum incubation temperature for thermophilic bacteria
	37	98	Incubation temperature for mesophilic bacteria
── 30 minutes ──	28	90	
1 hour ***Bacteria double every**	21	70	⎤ Incubation temperature for psychrophilic bacteria
2 hours	10	50	⎦
6 hours	4	40	⎤ Safe refrigerator storage temperature
── 20 hours ──	0	32	⎦
	-18	0	Maximum safe storage temperature for frozen chicken
	-29	-20	⎤ Quick frozen
	-40	-40	⎦
	-51	-60	⎤ All water in chicken freezes
	-73	-100	⎦

*Spoilage zone — bacteria and food toxins develop fast.
Source: USA Poultry & Egg Export Council.

Storing Poultry

Cleaning a chicken only when you plan to cook one is not an economical use of time. Due to the mess involved, butchering several birds at once makes more sense. Freshly butchered chicken meat will keep for up to five days in a refrigerator with a temperature between 29°F and 34°F (-1.6° – 1.1°C). To store it longer, either freeze it or can it within three days of butchering.

Freezing

Freezing preserves the quality of fresh meat and retards growth of bacteria. Since freezing does not *stop* bacterial growth, strict cleanliness is essential in all aspects of butchering and packing.

The quality of frozen chicken depends on:

◆ how fresh it is when you freeze it
◆ the temperature under which it remains frozen
◆ the length of storage time.

Chicken may be frozen raw or cooked, whole or in pieces. If you package cut up chicken, place just enough pieces in each packet for one meal. If you freeze whole birds, don't stuff them before freezing them. Dense stuffing slows freezing, giving bacteria more time to proliferate.

Whether you freeze birds whole or in pieces, trim away excess fat (since fat goes rancid fairly rapidly) and cut off sharp bones that may pierce the wrapping, exposing the meat to freezer burn. Freezer burn occurs when food is not adequately wrapped, allowing air to circulate over the exposed surface, removing moisture.

After trying all sorts of packing methods, I finally eliminated freezer burn by double-wrapping each packet, first in heavy foil or plastic wrap, then in waxed butcher paper sealed with freezer tape. If I'm packing whole birds, instead of butcher paper I use plastic bags designed for freezer use.

To remove air from a plastic bag, gather the opening of the bag and use a vacuum pump or vacuum cleaner to remove the air. (At one time, inhaling through a straw was recommended for drawing out air, but you run the risk of inhaling a tiny bit of foreign matter sucked from inside the bag.) With much of the air out, the bag will cling tightly to the meat. Twist the bag closed and apply a twist-tie.

An alternative to double-wrapping is to use cryo-vac bags, if you can find them. Cryo-vac is a kind of plastic that shrinks when dipped in boiling water.

On each packet, jot down the date (so you'll use the oldest first) and the contents (so you won't keep the freezer door open while you try to guess). Spread packets around in the freezer, leaving space between them for air to circulate until they freeze hard, which takes at least 12 hours. After the meat is frozen, you can stack it any way you like.

During freezing, ice crystals form from moisture in the meat. Quick freezing produces small ice crystals. Slow freezing causes big crystals that damage meat tissue. To ensure quick freezing, don't add more than 3 pounds of unfrozen meat per cubic foot (45 kg per cubic meter) of freezer capacity at a time.

Storing chicken in the freezer for any appreciable length of time is an option only if you have a dedicated freezer. The freezer compartment of most refrigerators isn't cold enough to hold meat for more than a couple of weeks.

To monitor the quality and storage time of your frozen chickens, keep a reliable thermometer in your freezer. Chicken stored at 0°F (–18°C) or lower retains its quality longer than meat stored at a higher temperature. Even though meat may remain cold and hard at temperatures above 0°F, it deteriorates more rapidly. If you have a good freezer that maintains a temperature of –10°F (–23°C), you can safely store chicken meat for up to a year.

Poultry Storage Guide

	Refrigerator (35–40°F) (2–4°C)	Freezer (0°F) (-18°C)
Raw		
Whole	5 days	12 months
Pieces	2 days	9 months
Giblets	1–2 days	3 months
Cooked		
Slices or pieces	1–2 days	1 month
Slices or pieces in gravy or broth	1–2 days	6 months
Casserole	1–2 days	6 months
Fried	1–2 days	4 months
Gravy or broth	1–2 days	3 months

Meat stored longer than the recommended time may develop an off flavor.

Freezing Cooked Chicken

Freezing fully cooked chicken is a handy way to have quick meals later on. If you freeze a roasted, stuffed bird, remove the stuffing and freeze it separately from the meat. Frozen cooked meat eventually takes on a rancid taste, but packing cooked meat in broth or gravy keeps air out and lengthens the storage time.

Cooked meat in a liquid or semi-liquid base may be frozen in heat-sealed boilable plastic bags. To ensure a complete seal, cool the food before filling the bags, and fill them so they're no more than 1-inch (2.5 cm) thick laid flat. When you're ready to reheat the chicken in a pouch, drop the whole pouch into boiling water.

To freeze a chicken casserole, cook the dish in an ovenproof or microwaveable container lined with foil. Cool the casserole, cover it, and freeze it — container and all. When the food is fully frozen, remove the casserole from its dish, double wrap it, and freeze it. When you're ready to reheat it, unwrap the casserole and drop it back into the original container.

Any time you freeze cooked food, cool it first so it won't heat up the inside of your freezer. Pack the cooked food for freezing, then let it cool at room temperature for 30 minutes before popping it into the freezer.

Power Failure

If your power goes out, according to the USDA you can safely refreeze chicken if:

◆ it still contains ice crystals *or*
◆ the temperature has been above freezing (but below 40°F, or 4°C) for no more than two days.

When the power goes out, the first thing to do is avoid opening the freezer, even if it means preparing something other than the meal you had planned. If you keep the door closed, a loaded freezer will stay cold enough to preserve chicken and other foods for one or two days, depending on its size. Even a partially loaded freezer should keep meat frozen for a day. You can prolong thawing within the freezer by wrapping the unit in comforters, quilted furniture pads, bed pads, or anything else that's large and thick.

If you live in an area where power outages are frequent, try in advance to find a place where you can get dry ice in a hurry. Dry ice is carbon dioxide, solidified at a temperature of -220°F (-140°C). As it warms, it turns into a gas that evaporates, leaving no puddle like "wet" ice does. A 50-pound block will keep a 20-cubic-foot freezer going for up to four days if it's fully loaded, three days if it's at least half full. To avoid burning your fingers when handling dry ice, wear gloves and leave the ice in its brown paper wrapper.

Although refrozen chicken is safe to eat, it won't taste as good as meat that hasn't been thawed and refrozen. To preserve remaining flavor, cook the chicken before refreezing it. Even raw meat that's too far thawed to refreeze, but is still safe to eat, may be cooked and refrozen. Unfortunately, cooked meat that has thawed should not be refrozen, but should be used within two days.

Thawing

Frozen chicken pieces may be cooked without being thawed, but they'll cook faster if thawed first. A whole chicken should always be thawed before being roasted. Chicken may be safely thawed in one of four ways:

In the refrigerator a 4-pound (1.8 kg) chicken will thaw in about a day. This is the safest method, since you don't run the risk of forgetting about the meat and letting it get too warm for too long. Put the package of frozen chicken on a plate or tray to catch drips and place the plate in the refrigerator until the meat is pliable.

Thawing Times in Refrigerator

Size of Pieces	Hours
cut up	4–9
halved	3–9
whole, under 4 pounds	12–16
whole, 4 pounds and more	24–36

In a cool room a 4-pound (1.8 kg) chicken will thaw in about 12 hours. As long as the temperature is no more than 70°F (21°C), wrap the frozen package in several thicknesses of paper and leave it out until the meat becomes pliable.

Thawing chicken on a counter in a warm room isn't wise because of the danger that you might forget the meat, letting it get too warm for too long. In a really warm room, the outside layer may thaw and start to spoil while the inside is still frozen.

In cold water a 4-pound (1.8 kg) chicken will thaw in two hours or less. Seal the frozen packet in a plastic bag and submerge it in cold water, changing the water often, until the meat is pliable.

In a microwave oven frozen chicken may be thawed in a matter of minutes. Follow your microwave manufacturer's directions.

If you start to thaw chicken in the refrigerator or in a cool room, and it isn't completely thawed by the time you're ready to cook it, speed things along by putting the packet in cold water or in the microwave. Another way to shorten thawing time is by freezing pieces with waxed paper or freezer wrap between them, so you can easily separate the pieces for thawing.

Don't be alarmed if frozen chicken looks dark near the bone after it's cooked. Darkening is a reaction to slow freezing that normally occurs in home-frozen chicken.

Canning

To safely can chicken you need a pressure canner; for pint and half-pint jars you may use a pressure saucepan. No method that processes without pressure is safe for meat. Chicken, like other meat, may contain bacteria that cause botulism, a form of food poisoning. The bacteria are destroyed by processing the meat at 240°F (116°C) for a specific length of time that depends on the volume. If the temperature is lower or the time is shorter, the risk of botulism occurs.

If you are not familiar with the use of a pressure pot, do not attempt to can chicken based on the suggestions offered here. Read the manual that came with the canner, consult a good canning guidebook, or get information from your state or county Extension home economist.

A pressure canner operated at sea level at 10 pounds of pressure reaches a temperature of 240°F (116°C). If you live above sea level, adjust the pressure and timing for your altitude, as specified in your manual. If you put up pints or half-pints in a pressure saucepan, add 20 minutes to any specified processing time.

Chicken meat is easier to handle and will be more tender if it has been chilled before being canned, but there's nothing wrong with canning it as soon as the body heat is gone. Remove as much fat as possible. Do not can excessively fatty pieces. Add salt for flavor, if you wish, or leave it out, if you prefer — salt does nothing to preserve the meat. Work as quickly as possible and process the jars as soon as they are filled.

Chicken Soup

At our house, whenever anyone feels under the weather, we warm up a jar of homemade chicken soup. As a cold remedy, chicken soup has the blessings of the doctors at the Mayo Clinic. Maybe the vapors shrink swollen nasal passages. Maybe the garlic pickles microbes. Maybe cysteine, an amino acid in chicken broth, eases congestion. Who cares, as long as it works.

We always have chicken soup on hand because, whenever we butcher, we cook up the bony parts (wings, backs, necks), debone them, and make at least one 7-quart cannerful of soup. Any remaining meat is hot-packed for use in salads and sandwiches.

Homemade Chicken Soup

My husband, Allan, devised this delicious recipe that makes a 7-jar canner load of chicken soup, plus 8 half-pint jars of hot-pack meat: Pressure cook the bony parts of 10 chickens, adding 8 quarts of water and processing the parts at 15 pounds for 25 minutes (or simmer in a kettle until the meat falls off the bone). Debone.

In each of seven one-quart jars, place (in this order) —

½	cup thin egg noodles, uncooked
⅓	cup chopped celery
⅓	cup chopped onion
¼	cup chopped carrots
1	clove garlic, crushed
¼	cup cooked deboned chicken
1	teaspoon salt
⅛	teaspoon pepper
3½	cup strained hot broth

Leave 1 inch of head space. Clean the jar rims well and screw on clean lids. At sea level, process quart jars at 11 pounds of pressure for 45 minutes.

If you have broth left over after hot-packing any remaining meat, put it up in quart jars, processed for 25 minutes at 10 pounds of pressure. Or freeze it in ice-cube trays for use in gravies and to flavor vegetables. After the cubes have frozen, remove them from the trays, pack them in freezer containers, and return them to the freezer.

Hot Pack

Chicken may be hot-packed with or without the bones, but deboned canned chicken is easier to use, and you can get more meat into each jar. Deboning lets you make good use of meat from bony parts to make salads and sandwiches.

Simmer the bony pieces, covered in water, just until the meat starts to fall off the bone. Remove all the bones and skin, and pack the meat loosely into clean glass jars. Keep the broth simmering.

If you wish, add salt: ¼ teaspoon per half-pint, ½ teaspoon per pint, 1 teaspoon per quart. Cover the meat with simmering broth, leaving 1 inch of head space. Wipe the jar rims to rid them of fat and meat particles. Seal jars with clean lids. At sea level, process jars at 11 pounds:

> **half-pints, 60 minutes**
> **pints, 75 minutes**
> **quarts, 90 minutes**

Raw Pack

If you wish to can chicken pieces bones and all, raw packing is easier than hot packing. For a raw pack, canning the meaty pieces (thighs, breasts, and drumsticks) makes more sense. Since the breastbone and drumstick take up lots of room, at least bone the breasts and saw drumsticks short. Trim off any fat and pack pieces loosely into quart jars.

Place thighs and drumsticks with their skin next to the glass. Fit breasts into the center. Use smaller pieces to fill up the remaining space. Leave 1 inch of head space at the top of the jar. If you wish, add 1 teaspoon of salt. You need not add liquid — raw meat generates its own juice while it cooks.

Wipe the jar rims to rid them of fat and meat particles. Seal the jars with clean lids. At sea level, process jars at 11 pounds for 80 minutes.

Canned Meat Safety

To ensure the safety of canned meat:

- keep all equipment clean
- meticulously follow processing times and temperatures
- let jars cool before storing them in a cool place
- boil home-canned meat 20 minutes before serving it.

If you wish to use canned chicken for cold salads or sandwiches, chill the meat in the refrigerate after it's been boiled.

Signs of spoilage include:

- bulging or leaking jar lids
- gas bubbles inside a jar
- liquid spurting out when a jar is opened
- off colors or off odors

The best way to find out if canned meat is safe to eat is to boil it. Boiling brings out the characteristic smell of spoilage. If canned meat doesn't look or smell right, dispose of it as you would any toxic substance.

Marketing Meat Birds

Selling meat birds can be profitable. It can also be economically risky. For starters, you'll have a hard time competing with low market prices unless you find customers willing to pay a premium for homegrown or "organic" meat. For another thing, many customers are not prepared for the differences in taste, texture, and appearance between homegrown and commercially grown meat. Selling meat birds therefore requires a certain amount of consumer education.

To add to your woes, you may need expensive state-approved facilities and a license to sell dressed birds, although in some states small producers enjoy exemptions. Your state poultry specialist can fill you in on the details. You may be able to get around dressed-bird laws by using a little creativity, such as selling live birds and giving buyers the option of plucking their own or having you pick them for free.

Marketing possibilities available in some areas include selling live birds to local butchers, custom pickers, or processing plants. Processing plants, which generally want large numbers of birds, operate under various kinds of contracts. At one end of the spectrum, you may be required to purchase all your chicks and feed with the understanding that you'll sell back the finished birds. At the other end, chicks and feed may be furnished without charge and you'll be paid for your labor and expenses. Be sure to read the contract's fine print, including who's responsible for chicks that die or fail to measure up to market quality.

Unless you're willing literally to eat your profits, determine who your customers will be before embarking on a meat-bird marketing venture.

Organic Certification

If you're going to sell meat birds, you may wish to obtain organic certification. Certification is basically a marketing tool that reassures customers of the quality of your product and justifies your higher prices compared to factory-farmed chicken.

Until the USDA finalizes federal standards for organic poultry production, some groups have been providing private certification. Various certification groups work on local, regional, national, and international levels, but not all of them certify animal products. You'll need to find the group that best suits your needs; then determine whether or not you meet their criteria.

The never-ending stream of paperwork required for certification discourages a lot of people right from the start. First there's an application, which in itself can be a formidable document. Then there's an interminable amount of record keeping, detailing every step of production. Your operation will be inspected before your application is approved and every year thereafter. Finally, you'll have to pay certain fees, which may include a flat fee, an inspection fee, and a percentage of your sales volume.

For up-to-date details on organic certification and a list of programs, obtain the information packet "Organic Certification" from ATTRA (Appropriate Technology Transfer for Rural Areas, listed in the appendix under *Information Packages*).

CHAPTER 6

LAYER MANAGEMENT

EACH PULLET starts life carrying the beginnings of as many as 4,000 undeveloped eggs inside her body. Rare is the hen that lays more than about 10 percent of the total number within her capability. By keeping your hens healthy and happy, you encourage them to lay the greatest possible number of eggs.

Laying Breeds

Your first decision in establishing a layer flock is what kind of hens to raise. You have two basic choices:

- ◆ a commercially developed layer strain
- ◆ an old-fashioned laying or dual-purpose breed.

Most commercial strains are genetically based on the Leghorn. Leghorns are bred to start laying at an early age and to lay a large number of eggs in a short time. Their high production rate can be accounted for, in good part, because they are small-bodied and therefore able to channel fuel (in the form of feed) into egg production rather than into body maintenance.

A Leghorn begins laying at the age of 18 to 22 weeks and will lay between 250 and 280 eggs during her first year. Individual hens may exceed the average — the greatest number ever laid, according to Guiness, was 371 eggs produced in 364 days by a prolific Leghorn in 1979 at the University of Missouri.

Most commercial layers are not hybrids in the strictest sense, since they're bred by crossing different strains of Leghorn rather than different breeds. One that truly is a hybrid is the Red Sex Link, a cross between a white Leghorn hen and a Rhode Island Red cock. The Red Sex Link, developed commercially to produce brown eggs, lays 250 to 260 eggs per year.

SHANK'S HATCHERY

Black Sex Links are but one strain among many commercial brown-egg hybrids.

The Black Sex Link is another brown-egg hybrid, but one that is not Leghorn-based, being a cross between a barred Plymouth Rock hen and a Rhode Island Red cock. The Black Sex Link lays somewhat fewer eggs than the Red — about 240 per year — but her eggs are larger. The Black Sex Link is gentler than the Red and quicker to mature, beginning to lay at about 16 weeks of age. Although she eats a little more than the Red, at the end of her productive life she weighs 5 pounds (2.25 kg) — a good ¾ pound (0.33 kg) more than the Red — making the Black something of a dual-purpose hybrid.

Since purebreds characteristically put more energy into muscle mass than do commercial layer strains, most are considered dual-purpose birds. Of these, the Rhode Island Red is indisputably the most popular for backyard flocks. A purebred hen starts laying at the age of 24 to 26 weeks. You won't get as many eggs from the best purebred as from the best commercial strain, and you won't get as many eggs from a purebred strain bred for show as from a strain of the same breed bred for production.

Shell Color

As a general rule, hens with white ear lobes lay white eggs and hens with red ear lobes lay brown eggs. Exceptions to the red-ear-lobe rule are Crevecoeur, Dorking, Lamona, Redcap, and Sumatra, which lay white-shelled eggs, and Araucana and Ameraucana, which lay blue-green eggs.

All Mediterraneans lay white eggs. All Asiatics and most Americans (except Holland and Lamona) lay brown eggs. Other classes are a mixed bag as to white, brown, tinted (light brown to the point of pinkish), or blue-green. These colors are the result of pigments in the outer layer of the shell. Although variations in hue exist within each breed or strain, an individual hen characteristically lays eggs of a specific color.

Shell Color by Breed

White–Egg Layers		Brown–Egg Layers	
Ancona*#	Minorca	Aseel	Java
Blue Andalusian	Modern Game#	Australorp	Jersey Giant
Campine*#	Old English#	Black Sex Link##	Langshan
Catalana#	Polish	Brahma	Malay
Crevecoeur	Redcap	Buckeye	Naked Neck
Dorking	Sicilian Buttercup	Chantecler	(Turken)
Hamburg*#	Silkie	Cochin	New Hampshire
Holland	Sultan	Cornish	Orpington
Houdan	Sumatra#	Delaware	Plymouth Rock
LaFleche	White Faced Black	Dominique	Red Sex Link##
Lakenvelder#	Spanish	Faverolle	Rhode Island Red
Lamona	Yokohama	Frizzle**	Sussex
Leghorn		Hubbard Golden	Wyandotte
		Comet##	

*Lays smallish eggs.
#Shells may be lightly tinted.
**Frizzledness is a genetic condition that can occur in white-egg-laying breeds.
##Hybrid.

Rate of Lay

All pullets lay small eggs when they first start out, and they lay only one egg every three or four days. By the time a hen is 30 weeks old, her eggs will reach their normal size and she should lay at least two eggs every three days. A good laying hen produces about 20 dozen eggs in her first year. At about 18 months, she'll take a break to molt. After the molt, she'll lay bigger eggs than before, but fewer of them. During her second year, she'll average between 16 and 18 dozen.

A hen's rate of lay is affected by a number of external factors including temperature and light. Hens lay best when the temperature is between 45°F and 80°F (7°–27°C). When the weather gets much colder or much warmer, production slows down.

All hens stop laying in winter, not because the weather turns cold, but because daylight hours are shorter in winter than in summer. When the number of daylight hours falls below 14, hens may stop laying until spring.

A healthy hen should lay for a good 10 to 12 years — occasionally you'll hear of a biddy laying to the ripe old age of 20 — but most layers don't manage to live that long. Instead, at the tender age of one or two years they succumb to the firing squad or a falling ax, to be replaced by fresh young pullets.

Flock Replacement

Pullets generally reach peak production at 30 to 34 weeks of age. From then on, production declines approximately ½ percent per week until the birds molt or are replaced. Some producers molt their birds (discussed later in this chapter) at the age of about 60 weeks to induce a second round of laying. Others raise a new batch of pullets to replace the older hens when they reach 72 weeks of age.

Keeping layers for a second year has quite a few disadvantages:

◆ As hens get older and their production declines, the shells of their eggs get rougher and weaker, the whites get thinner, and the yolk membranes become so weak they break when the eggs are opened into a pan.

◆ As production declines, the cost of feeding older hens becomes greater than the value of their eggs.

◆ If you're selling eggs, production by older hens may fall below the numbers you need to meet the demands of established customers.

◆ While you may find a market for 1-year-old layers, by the end of their second year hens lose nearly all of their laying value and are good for little more than stewing.

◆ The older a chicken gets, the more likely it is to experience disease complications.

Your decision on whether to replace your layers every year or every two years will depend, in part, on whether you keep hybrids or purebreds. Commercially developed hybrids still produce fairly well during their second year, while purebreds peter out somewhat faster. Dual-purpose hens such as New Hamps that are more suitable for meat than for eggs tend to run to fat and do not lay well beyond the age of 9 to 10 months. On the other hand, purebreds that are better known for eggs than for meat may produce at a low but steady rate for years.

Your decision may be influenced by whether or not your hens are range fed. Pastured hens are considerably cheaper to feed than confined hens. Given the lower feed costs, you may not mind getting fewer eggs.

Your decision will be strongly influenced by the cost of starting new pullets compared to the cost of keeping the old flock. If you must purchase replacement chicks (as you would have to do if you raise hybrids), you may find it more economical to keep hens that are already laying than to purchase and raise a new batch. If you keep purebreds, hatching your own chicks will save you the considerable cost of purchasing replacements. By the same token, if you can realize any salvage value from the old hens (as meat birds or as layers sold to someone with less demanding needs than yours), you have another way to off-set the cost of raising new pullets.

In any case, do not be tempted to boost egg production by periodically bringing pullets into your established flock. Constantly introducing new birds disrupts the peck order, resulting in fighting and sometimes cannibalism. The ensuing stress can reduce the rate of production of pullets and hens alike. Intro-ducing new birds also increases the chance of introducing disease. Younger birds may bring in a disease, or they may catch something from the older birds. Also, as the older hens continue to age, their reduced productivity will depress the flock's overall average.

Feeding Pullets

Layer pullets should grow slowly so they'll be fully developed by the time they come into lay. Otherwise they may start laying early, before their bodies are ready for production, or they may get too fat. As a result, they'll tend to lay fewer, smaller eggs and will be more likely to prolapse (discussed in a moment).

Restricting the feed intake of pullets past the age of 4 weeks slows growth and delays the onset of laying, causing them to do better as layers and/or breeders. But because they spend less time eating, they have more time to get bored and nervous, and may pull each other's feathers or otherwise engage in cannibalism.

To prevent these problems, do not limit the total amount you feed. Instead, when layer pullets reach 8 weeks of age, gradually add oats until you achieve a balance of about 50/50, and continue feeding free choice.

When your pullets come into lay, and as they reach peak production, their need for dietary protein will go up. At about 18 weeks (20 weeks for heavier breeds), gradually reduce the oats and switch from grower or developer ration to 18 percent lay ration. Do not switch to lay ration too soon — its high calcium content may interfere with bone formation, resulting in weak legs, kid-ney damage, and possibly death.

To ensure that your pullets get the nutrients their bodies need as they come into production, have their commercial feed completely switched to 18 percent lay ration by the time they start to lay. By then, any medicated feed you may have used must be discontinued to avoid problems such as abnormal eggs and drug residues in eggs.

Each commercial strain lightweight pullet will eat about 15 pounds of feed before she starts laying at about 20 weeks of age. Heavier pullets take a few weeks longer to reach laying age, during which they eat as much as double the feed. Age at lay, and the amount of feed needed to bring pullets to that point, will vary with the specific breed and strain you choose.

Pullet Problems

A pullet that starts laying at too young an age, that is too fat or unhealthy when she comes into production, or that lays unusually large eggs may experience egg binding or prolapse.

Egg Binding occurs when a too-large egg gets stuck just inside the vent. To dislodge the egg, lubricate a finger with mineral oil or KY Jelly and insert it into the vent. With your other hand, push gently against the hen's abdomen and work the egg toward the vent. If you can see the egg but can't get it out, puncture the shell and remove the egg in pieces (taking care not to injure the hen with a sharp shard). Rinse away remaining egg bits with hydrogen peroxide in a squirt bottle. If any tissue protrudes through the vent, treat the pullet as you would for prolapse.

Prolapse (also called "blowout" or "pickout") occurs when pink tissue just inside the vent gets pushed to the outside of the vent. The exposed tissue attracts other birds to pick, and the pullet eventually dies as a result of hemorrhage and shock. If you catch prolapse in time, you may be able to reverse the situation by applying a hemorrhoidal cream (such as Preparation H) and isolating the pullet until she heals.

Bleaching Sequence

The skin of a pullet contains a considerable amount of yellow pigment, obtained from green feeds and yellow corn. Instead of using this pigment to color her skin, a hen in lay uses it to color the yolks of her eggs. As time goes by and the hen's old yellow skin is replaced with new pale tissue, her body parts appear to bleach out. After six months of intense laying, a high producing yellow-skinned hen will be completely bleached.

Pigment leaves the body parts in a certain order, based on how rapidly the skin of each part is renewed — the more rapidly the tissue is replaced, the more quickly it appears to bleach out. When a hen stops laying, color returns to her skin in reverse order, approximately twice as fast as it disappeared. You can therefore estimate how long a yellow-skinned hen has been laying, or how long ago she stopped laying, by the color of her various body parts.

Bleaching Sequence

Body Part	Eggs Laid	Weeks*
1. Vent	0–10	1–2
2. Eye ring	8–12	2–2½
3. Ear lobe	10–15	2½–3
4. Beak	25–35	5–8
5. Bottom of feet	50–60	8
6. Front of shank	90–100	10
7. Hock	120–140	16–24

*Individual bleaching time depends on a hen's size, her state of health, her rate of production, and the amount of pigment in her rations.

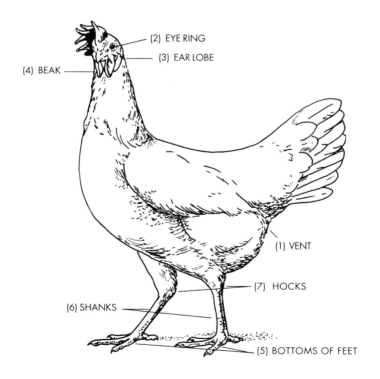

Pigment is "bleached" from the skin in this sequence: (1) vent, (2) eye ring, (3) ear lobe, (4) beak, (5) bottoms of feet, (6) shanks, and (7) hocks.

The first color loss is in the skin around the vent, which is renewed quite rapidly. Just a few days after a pullet starts to lay, her vent changes from yellow to pinkish, whitish, or bluish. Next to bleach is the eye ring. Within three weeks, the ear lobes of Mediterranean breeds bleach out; hens with red lobes do not lose ear color.

The beak's color fades from the corner outward toward the tip, with the lower beak fading faster than the upper beak. In breeds that characteristically have dark upper beaks, such as Rhode Island Red and New Hampshire, only the lower beak is a good indicator.

The best indicators of long-term production are the shanks, since they bleach last. Color loss starts at the bottoms of the feet, moves to the front of the shanks, and gradually works upward to the hocks.

Feeding Hens

Each lightweight layer eats about 4 pounds (1.8 kg) of feed for every dozen eggs she lays, which works out to between 4 and 4½ ounces (112–130 g) of feed per hen per day, or just under 2 pounds (0.9 kg) per hen per week. Dual-purpose hens eat a bit more (about ⅓ pound or 135 g per day), bantams a bit less. Feed your hens free choice, and they'll eat as much as they need.

Since chickens eat to meet their energy needs, expect your layers to eat less in summer than in winter, when they need extra energy to stay warm. If their summer ration contains the same amount of protein as their winter ration, they'll get less total protein in summer and therefore won't lay as well. In warm climates, in addition to regular 16 percent lay ration, some feed stores offer a ration containing 18 percent or more protein for use when high temperatures cause hens to eat too little.

Lay ration is designed as a complete feed that provides all a hen's nutritional needs. It comes either crumbled or in pellet form. Pellets contain the same ingredients as crumbles, pressed into long thin tubes and cut into little pieces. Hens fed pellets waste less than those fed crumbles.

Supplementing rations with table scraps and surplus milk products may reduce the cost of egg production, but it can also reduce production. To keep up production, offer such treats in amounts no more than your hens will polish off in 10 minutes, once a day.

If you have access to an inexpensive source of grain, as well as a source for lay ration containing 20 percent protein (or higher), you may be able to save money by combining the grain with the pellets in the ratio of 40 to 60. The high protein lay ration will offset the low protein grain to maintain the nutritional balance your hens need.

If you use a lay ration containing less than 20 percent protein, restrict the scratch allotment for confined layers to no more than 5 percent of their total diet, or about as much as they will eat in 20 minutes. Otherwise their energy-protein balance will be thrown off, they'll get fat, and they will lay poorly. Range-fed hens aren't quite as touchy when it comes to grain, since they burn off the extra energy while foraging.

Range Feeding Layers

Range feeding your layer flock will reduce your feed bill by up to 30 percent but will require a fair amount of land. Although you can range as many as 200 hens on an acre, you need to move the flock every three to four days, so the total amount of land required is more like 50 acres per 100 hens. Range-fed layers are therefore best worked into a pasture rotation scheme involving other livestock such as goats, sheep, or cattle.

The range house should provide at least 1 square foot (940 sq cm) of space per hen and should have nests that are equally accessible to birds from the inside and to you from the outside. Although you'll need to move the house a couple of times a week, you can return it to a previous site that has been rested for at least a month. To optimize feed-cost savings, move the henhouse as soon as you notice ration consumption starting to rise.

Move a portable henhouse as soon as you notice an increase in commercial ration consumption.

Supplemental Calcium

Lay rations contain 2.5 to 3.5 percent calcium, enough to meet a pullet's needs if she eats nothing but commercial feed. Older hens, and all hens that eat grain or grass in addition to commercial rations, need supplemental calcium in the form of oyster shell or soluble calcium grit offered free choice.

If you want to go through the trouble, you can recycle the shells from your hens' own eggs. Wash the shells, dry them, crush them, and feed them back to your hens as a calcium source. Feed only shells that have been crushed, or you may encourage hens to eat their own eggs. Store and feed only dry shells, since wet shells attract and harbor bacteria.

Egg Formation

A pullet starts life with two ovaries, but as she matures the right ovary atrophies and only the left one develops to maturity. The functioning ovary contains the 4,000 or so undeveloped yolks, or ova, the pullet was born with. If you have occasion to examine a hen's innards, you'll find the ova in a clump along her backbone, approximately halfway between her neck and tail.

When a pullet is ready to lay, one by one the ova mature, so that at any given time her body contains ova at various stages of development. Approximately every 25 hours, one ovum acquires enough layers of yolk to be released into the oviduct, a process called *ovulation*. Ovulation usually occurs within an hour after the previous finished egg was laid. During the yolk's journey through the oviduct, it is fertilized (if sperm are present), encased in various layers of albumen, or egg white, wrapped in membranes, and sealed in a shell.

The whole process takes about 25 hours, causing a hen to lay her egg about an hour later each day. Since a hen doesn't like to lay in the evening, eventually she'll skip a day and start a new laying cycle the following morning. The group of eggs laid within one cycle is called a "clutch." Some hens take more time than normal (say, 26 hours) between eggs, and therefore lay fewer eggs per clutch than a hen that lays every 25 hours. Conversely, some hens lay closer to every 24 hours, and so lay more eggs per clutch. Production hens are bred to have the shortest possible interval between eggs so they will lay as many eggs as possible per clutch. The best heavy breed hens in peak production lay 40 eggs in a clutch; a superior Leghorn lays 80.

The size of a hen's eggs depends on her breed and age. As a hen gets older, her eggs get bigger. Egg size is also influenced by a hen's weight. Underweight pullets lay small eggs. The first eggs of any pullet are usually quite small, but their size continues to increase until the bird is about 12 months old. Pullets that start laying during summer usually lay smaller eggs than pullets that start laying during the cooler months of fall or winter. Hens of any age and weight may temporarily lay small eggs if they're suffering from stress induced by heat, crowding, or poor nutrition, including inadequate protein or salt.

Eggs are classified by size according to their minimum weight per dozen. Within a flock of a given breed or strain, the majority of eggs should be of one size, with a few eggs ranging just above or below the majority size.

Egg Sizes

Size	Weight per Dozen	
Peewee	15 ounces	428 grams
Small	18	512
Medium	21	596
Large	24	680
Extra large	27	764
Jumbo	30	841

Egg Problems

The number of eggs a hen lays and the size, shape, and internal quality of the eggs — as well as shell color, texture, and strength — can be affected by a variety of things, including environmental stress, improper nutrition, medications, vaccinations, parasites, and disease.

Floor eggs, or eggs laid on the floor rather than in nests, are the first problem you are likely to encounter in a flock of pullets just starting to lay. Floor eggs get dirty or cracked, making them unsafe to eat. A cracked egg is easily broken, encouraging birds to sample the contents, develop a taste for eggs, and thereafter become egg eaters. To minimize floor eggs, put nests out early so your pullets will get used to them before they start laying. Place the nests low to the ground until most of the pullets are using them, then raise the nests 18 to 20 inches (45–50 cm) to discourage birds from entering nests for reasons other than to lay.

If pullets continue laying on the floor, perhaps you have too few nests — provide at least one for every four layers. Or perhaps the nests get too much light, causing pullets to seek out darker corners for laying. Since the primary purpose of laying eggs is to produce chicks, layers have a deep-seated instinct to deposit their eggs in dark, protected places. A properly designed nest offers them just such a place.

A nest egg, or fake egg, left in each nest shows pullets the proper place to lay. When a hen sees an egg already in the nest, she says to herself, "Ah-ha, this must be a safe place for my own egg." Old golf balls make good nest eggs. You can find toy eggs in stores around Easter; wooden eggs, available year-round at hobby shops, are better than air-filled plastic eggs because they can't as easily be bounced out of the nest. It's an old wives' tale that nest eggs make hens lay more. By encouraging hens to lay in nests rather than, say, in the bushes, nest eggs help you find more of the eggs a hen does lay.

Dirty eggs in the nest are usually the result of layers tracking mud or muck on their feet. To keep eggs from getting dirty, clean up the source of mud — most often a muddy entry or damp ground around a leaky waterer — and take measures to ensure that the condition doesn't reoccur. You may also have to redesign your nests to discourage mud tracking. Eggs in nests located on or near the floor are more likely to get dirty than eggs in nests raised above the floor, especially if layers must get to raised nests only by hopping up on a perch or series of perches.

Eggs will get dirty, of course, if nest litter is not changed often or if the nests themselves collect rainwater. The latter can be a problem if a leaky roof covers nests designed for egg collection from outside the coop.

Occasionally eggs get dirty when birds low in the peck order spend a lot of time hiding in nests, soiling previously deposited eggs. Avoid crowding your flock and provide enough environmental variety to allow timid birds to get away from bullies.

Bloody shells sometimes appear when pullets start laying before their bodies are ready, causing tissue to tear. Other reasons for blood on shells include excess protein in the lay ration and coccidiosis, a disease that causes intestinal bleeding. Cocci does not often infect mature birds, but if it does you'll likely find bloody droppings as well as bloody shells.

Chalky or glassy shells occasionally appear due to the malfunction of a hen's shell-making process. Such an egg is less porous than a normal egg and will not hatch, but is perfectly safe to eat.

Odd-shaped or wrinkled eggs may be laid if a hen has been handled roughly or if for some reason her ovary releases two yolks within a few hours of each other, causing them to move through the oviduct close together. The second egg will have a thin, wrinkled shell that's flat toward the pointed end. If it bumps against the first egg, the shell may crack and mend back together before the egg is laid, causing a wrinkle.

Weird-looking eggs may be laid by old hens or by maturing pullets that have been vaccinated for a respiratory disease. They may also result from a disease itself, such as infectious bronchitis. Occasional variations in shape, which are sometimes seasonal, are normal. Since egg shape is inherited, expect to see family similarities.

Thin shells may cover a pullet's first few eggs or the eggs of a hen that's getting on in age. In the former case, thin shells occur because the pullet isn't yet fully geared up for egg production. In the latter case, the same amount of shell material that once covered a small egg must now cover the larger egg laid by the older hen, stretching the shell into a thinner layer.

Shells are generally thicker and stronger in winter but thinner in warm weather, when hens pant. Panting cools a bird by evaporating body water, which in turn reduces carbon dioxide in the body, upsetting the bird's pH balance and causing a reduction in calcium mobilization. The result is eggs that are thin-shelled.

Thin shells may be due to a hereditary defect, imbalanced rations (too little calcium or too much phosphorus), or disease — the most likely culprit being infectious bronchitis.

Soft or missing shells occur when a hen's shell-forming mechanism malfunctions or for some reason one of her eggs is rushed through and laid prematurely. Since the shell forms just before an egg is laid, stress induced by fright or excitement can cause a hen to expel an egg before the shell is finished.

A nutritional deficiency, especially of vitamin D or calcium, can cause soft shells. A laying hen's calcium needs are increased by age and by warm weather (when hens eat less and therefore get less calcium from their rations). Appropriate nutritional supplements include free-choice limestone or ground oyster shell, and vitamin AD&E powder added to drinking water three times a week.

Soft shells that are laid when production peaks in spring, and occasional soft or missing shells, are nothing to worry about. If they persist, however, they may be a sign of serious disease, especially infectious bronchitis, Newcastle disease, or infectious laryngotracheitis, all of which are accompanied by a drop in production.

Broken shells often result when thin or soft shells become damaged after the eggs are laid. Even sound eggs may get broken in a nest that's so low to the ground that chickens are attracted frequently to scratch or peck in them. Hens and cocks may deliberately break and eat eggs if they are inadequately fed or are bored. Boredom may result from crowding or from rations that allow chickens to satisfy their nutritional needs too quickly, leaving them with nothing to do. If your coop is small and well lighted, discourage nonlaying activity in nests by hanging curtains in front to darken them. To allay a hen's suspicions about entering a curtained nest, either cut each curtain into hanging strips or temporarily pin up one corner.

Hens may also break eggs inadvertently. Accidents may occur if nests contain insufficient litter, if eggs are collected infrequently enough to pile up in nests, or if nests are so few that two or more hens must lay in the same nest at the same time. Sometimes timid birds seek refuge by hiding in nests.

Too few eggs being laid can result from so many different causes you practically have to be Sherlock Holmes to determine the reason. For starters, you may be raising the wrong hens. Although production varies between individuals, strains, and breeds, if you want lots of eggs, you need a flock of hens that have been developed for production.

Even if you have the "right" breed, you'll get few eggs if your hens are old. Most hens lay best during their first year, although a really good layer may do well for two years or more.

During the molt, laying strains slow down in production, and purebreds may stop altogether. Low production as a result of an out-of-season molt is a sure sign of stress. Stress itself, with or without an accompanying molt, can cause hens to slow down or stop laying. A list of common stress situations appears on page 241.

Improper nutrition can cause a drop in production. Hens may get too little feed or may be fed rations containing too little carbohydrate, protein, or calcium. Imbalanced rations often result from feeding hens too much scratch and from failure to offer a free-choice calcium supplement when the diet includes grain or pasture. Low temperatures increase a chicken's requirement for carbohydrates, and unless rations are adjusted accordingly, low production may result. Dehydration due to lack of water for even a few hours can cause hens to stop laying for days or weeks.

You may get too few eggs if one of your hens hides her eggs where you can't find them or if she lays her eggs and then turns around and eats them. (Egg eating, a form of cannibalism, is discussed on page 282). An egg eater won't necessarily come from within your flock, but may be a wily predator (as discussed under "Predator Control" on page 263). If you can find no other cause for a drop in production, your hens may be coming down with a disease (see "Disease and Egg Laying" on page 271).

Controlled Lighting

In the natural course of events, chicks hatch in spring, when daylight hours are increasing, and mature during summer and autumn, when daylight hours are decreasing. The following spring, when day length once again begins to increase, they start a new reproductive cycle. Hens continue to lay until either the number of light hours per day or the degree of light intensity signals the end of the reproductive cycle.

To get eggs during winter, you have to trick your hens into thinking the season is right for reproduction. Do so by using artificial light to compensate for decreasing amounts of natural light. The farther you live from the equator, where day length is constant, the bigger your seasonal swings in increasing and decreasing day length will be.

Start augmenting natural light when day length approaches 15 hours, which in most parts of the United States occurs in September. Continue the lighting program throughout the winter and into spring, until natural daylight is back up to 15 hours per day.

If you forget to turn the lights on just one day, your hens may go into a molt and stop laying. To protect yourself from your own forgetfulness, use a timer switch. Set the timer to slightly overlap natural light. Add the light hours in the morning, rather than in the evening, so your hens won't suddenly get caught in the dark before they're settled on the roost.

A timer has to be constantly adjusted as the number of daylight hours change and — unless it's connected to an uninterrupted power source (UPS) — must be readjust every time the power goes out. Many people find it more convenient to leave lights on all the time. Constant lighting has its down side, though — it encourages hens to spend more time indoors during the day stirring up litter dust, scratching in nests, and otherwise engaging in mischief.

In an effort to decrease the already minimal cost of lighting, you might be tempted to install fluorescent lights. Although fluorescent tubes are cheaper to run than incandescent bulbs, they're more expensive to install, touchier to operate in the dusty henhouse environment, and more difficult to regulate. To adjust the light intensity of fluorescent lights you have to change the entire fixture; with incandescent lights you just switch to a bulb of different wattage.

If your coop is not outfitted with electricity, you can provide lighting with 12-volt RV bulbs, powered by a car battery connected to a solar recharger. Since a timer would drain too much power from your battery, 12-volt lights must either be left on all the time or manually turned on and off.

One 60-watt bulb, 7 feet (210 cm) above the floor, provides enough light for about 200 square feet (18 sq m) of living space. Place the bulb in the center of the area to be lighted, preferably over feeders. For caged layers, use bulbs of smaller wattage and place them closer together (see "Controlled Lighting" chart). If your coop is so large that you need more than one fixture, the distance between them should be no more than 1.5 times their height above the birds. Arrange multiple fixtures to minimize shadows.

A reflector behind each light increases its intensity, allowing you to use less wattage than you would otherwise need. In the above example, reflectors would let you substitute 40-watt bulbs for the 60-watt bulbs. Dust accumulating on bulbs decreases their light intensity. To maintain the effectiveness of your controlled lighting program, dust bulbs weekly and replace any that burn out.

Maximum Height of Lights Above Birds

Watts	Age 12–21 weeks		Age 21+ weeks	
	Reflector	No reflector	Reflector	No reflector
15	5'	3.5'	3.5'	2.5'
25	6.5'	4.5'	4.5'	3'
40	9'	6.5'	6.5'	4.5'
60	14'	10'	10'	7'
75	15.5'	10.5'	10.5'	7.5'

Arrange fixtures to minimize shadows and space them no farther apart than 1.5 times their height above the birds.

Lighting Pullets

Light affects not only the production of hens, but also the sexual maturity of pullets — the age at which they begin laying, the number of eggs they lay, and the size of their eggs. Under normal circumstances, pullets mature during the season of decreasing day length. If you raise pullets in the off-season, increasing day length that normally triggers reproduction will speed up their sexual maturity, more so the closer they get to laying age. Pullets that start laying before their bodies are ready will lay smaller eggs and fewer of them, and are more likely to prolapse.

Pullets should be kept either on a constant 8- to 10-hour day or in decreasing light. Pullets hatched from April through July can be raised under natural light. Those hatched from August through March need controlled lighting to delay maturity. Consult an almanac to determine how long the sun will be up on days occurring 24 weeks from the date of hatch. Add six hours to that day length and start your pullet chicks under that amount of light (natural and artificial combined). Reduce the total lighting 15 minutes each week, bringing your pullets to a 14-hour day by the time they start to lay. When they reach 24 weeks of age, add 30 minutes per week for two weeks to increase total day length to 15 hours.

Lighting Reminder

Raise pullets in constant 8- to 10-hour days or in decreasing daylight.
Keep layers in constant 15-hour days or in increasing daylight.

Culling

Within every breed or strain, some hens lay better than others. Rigorously culling (or eliminating) inferior and unhealthy birds that don't lay well has several advantages:

- reduces feed costs
- removes potential sources for disease
- increases available living space for productive hens
- increases feed and water space for productive hens
- improves a flock's overall laying average

Culling is an ongoing process involving the removal of injured or sick birds whenever you spot them. It's also a periodic process of determining whether or not each hen is laying, then determining how long she has been in or out of production.

The first time to cull heavily is when your flock reaches peak production at about 30 weeks of age. A second culling time is toward the end of the first year of production — a good hen lays for at least 12 months; a lazy layer takes an early break.

The first step in culling is to look at each hen's overall carriage. A high producer is active and alert. A low producer tends to be lazy and listless. Next look at details:

Feathers of a good layer are worn, dirty, and broken. Poor layers look sleek and shiny — more like show birds than working gals.

Combs and wattles of a good layer are large, bright, and waxy. Candidates for culling have small combs and wattles.

The *vent* of a good layer is large, moist, and oval. Cull candidates have tight, dry, round vents.

The *abdomen* should feel soft, round, and pliable under your hand — never small and hard.

Pubic bones — the pair of pointy bones located between the keel and vent — should have enough room between them for two or three fingers; the space between the pubic bones and keel should accommodate three fingers. A lazy layer is tight and nonflexible in these two areas.

Of all these various indicators, the most reliable ones are the pubic bones and the vent.

Culling Checklist

Body Part	Good Layer	Poor Layer
Carriage	active and alert	lazy and listless
Eyes	bright and sparkling	dull and sunken
Comb and wattles	large and bright	small and pale
Shanks	thin and flat	round and full
Back	wide	narrow or tapered
Abdomen	deep and soft	shallow and hard
Pubic bones*	wide apart and flexible	tight and stiff
Vent*	large and moist	puckered and dry
Plumage	worn, dry, dirty	smooth, shiny, clean
Molt	late	early
Skin	bleached	yellow

*Most reliable indicators.

Good Laying Hen

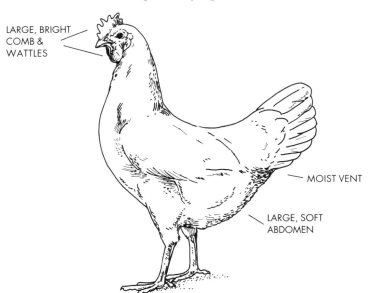

LARGE, BRIGHT
COMB &
WATTLES

MOIST VENT

LARGE, SOFT
ABDOMEN

In order to cull effectively, look at a bird's overall carriage and the specific qualities of the body parts listed above.

Molting

Short day lengths serve as a signal to birds that it's time to renew plumage in preparation for migration and the coming cold weather. Like all birds, chickens lose and replace their feathers at approximately one-year intervals. The process, called *molting*, takes place over a period of weeks, so a bird never looks completely naked (although one occasionally comes close).

Under natural circumstances, a chicken molts for 14 to 16 weeks during the late summer or early fall. The best layers molt late and fast; the poorest producers start early and molt slowly. Although no direct connection has been found between molting and laying, it makes sense that nutrients needed to produce eggs are channeled into plumage production. As a result, most pure-bred hens stop laying for about two months; commercially bred layers slow down in production but usually do not stop entirely.

Molting can also result from disease or stress such as chilling or going without water or feed. A stress-induced molt is usually partial and does not always cause a drop in production, while a normal full molt is usually accompanied by at least a production slowdown.

Molting Sequence

In full molt, feathers are renewed in sequence, starting with the head and gradually working toward the tail, with some areas molting simultaneously.

Controlled Molting

Since the molt is triggered by decreasing day length, layers kept in constant or increasing light do not have a normal molting pattern. As hens age, however, their eggs decline in quantity and quality. After 8 months or more of laying, a flock's production falls below the point of profitability. If keeping your old hens laying is less expensive than bringing in replacement pullets, consider deliberately sending your flock into a molt.

Causing hens to molt is called "controlled" or "forced" molting. Its purpose is to induce the whole flock to rest at the same time and come back into lay as a group. Hens that molt early enough to come back into production before fall lay better through the winter than late molters. So an additional purpose of controlled molting is to induce hens to molt when *you* want them to.

Not only will a flock's production average improve after the molt, but egg quality will go up, eggs will be larger, and the hens' feed efficiency will be better. On the other hand, your hens won't lay quite as well as they once did, and the quality of their eggs will decline faster than they did during the first lay.

Controlled molting is a tricky affair, since it involves inducing severe nutritional and environmental stress in order to stimulate more rapid than normal completion of the molt in 3 to 12 weeks. Leghorns, especially caged layers, respond best to controlled molting. Non-Leghorns and floor-raised birds require more severe stress. Take care — starving your hens and depriving them of water can backfire, causing deaths instead of rejuvenated laying.

The safest controlled-molting plan is to confine your flock where the ventilation is good and where you can limit lighting to 8 hours per day. Continue giving hens free-choice water, but discontinue lay ration. Instead, feed oats free choice, along with ½ pound (224 g) of scratch per dozen hens per day. Encourage rapid refeathering by including a vitamin/mineral supplement. After 2 weeks, gradually replace the oats with lay ration. At the same time, gradually increase lighting to 15 hours per day. Within about a month, your hens should be back into full production.

Culling by Molt

You can determine how long a hen has been molting, and how long she will continue to molt, by examining her primary, or flight, feathers — the largest feathers, running from the tip of the wing to the short axial feather that separates the primary from the secondary feathers. The ten primaries drop out at 2 week intervals and take approximately 6 weeks to regrow.

The primaries of a slow molter drop out one by one, requiring up to 24 weeks for the wing to fully refeather. A fast molter drops more than one feather at a time. Feathers that fall out as a group grow back as a group, letting the hen more quickly complete her molt and return to production. Culling slow molters that drop one feather at a time is one way you can keep your flock's production average up.

Culling by the Molt

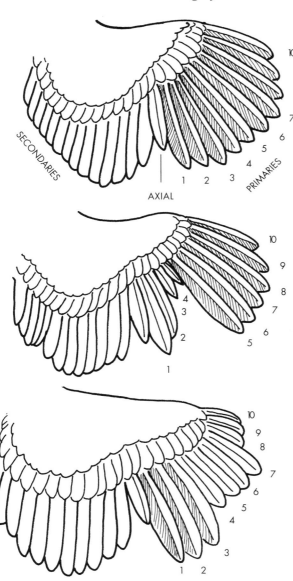

Fully feathered wing

Slow molt in which primaries drop out one by one. Feather #1 is fully regrown; #2 is 4 weeks old; #3 is 2 weeks old; #4 has just started. This wing has been molting for 6 weeks and will continue for 18 weeks more.

Fast molt in which primaries have dropped out in three groups. Feathers #1–3 are fully regrown; #4–7 are 4 weeks old; #8–10 are 2 weeks old. This wing has been molting for 6 weeks and will be finished in only 4 weeks more.

Egg Sales

If you're thinking of getting into the egg business, first define your market — to whom are you going to sell your eggs and how will you reach them. On a small scale, you may earn a dandy income selling eggs to neighbors and/or coworkers. For serious income, you will have to reach beyond those you know, perhaps working through natural-food stores, farmer's markets, and the like. One young fellow we know earns a nice little income peddling eggs to summer campers at a local beach.

If you go beyond the "We've got extra eggs, would you like to buy some?" stage, you'll have to define your product more formally. You might, for example, market "organic eggs from range-fed hens." Check local laws regarding claims you wish to make, conditions you have to meet in order to make those claims, and sales permits. Your county Extension agent or state poultry specialist should be able to fill you in.

Before getting into serious egg production, research your market so you'll know whether your customers will want white-shelled eggs or brown-shelled eggs. In most areas, brown eggs still command a higher price than white eggs, even though today's hybrid brown-egg layers are nearly as efficient as white-egg layers. On the other hand, brown eggs are more difficult than white eggs to candle for blood spots or meat spots (described on page 211), yet they tend to have a higher rate of these quality problems.

Pricing may require additional market research. Your customers may accept a fixed price or they may expect your price to fluctuate with market prices, which swing up and down with feed costs and consumer demand. When I lived in California, my customers were happy to pay a fixed price year-round. Here in Tennessee, some customers get indignant if I don't raise and lower prices to follow the market.

In establishing your price, take into consideration all your expenses, including not only production, but also applicable promotion, packaging, and delivery costs. The average market price is the *least* you should ask for your homegrown eggs. Since yours are fresher and tastier than store-bought eggs, you ought to get a few cents more.

Economic Efficiency

Industry uses several different methods to determine the economic efficiency of producing eggs. Since the cost of feed accounts for much of the expense of maintaining a layer flock, many efficiency indicators factor in feed use.

Feed cost is the total amount spent on feed during the production period. Naturally, the lower your feed cost, the better. Cost can be kept down by guarding against feed wastage and by seeking economical feed sources (including forage).

Feed as a percentage of total cost is the cost of feed consumed during a given production period divided by the total operating cost (including not only feed but also medications, litter, utilities, and other expenses). The higher this number is the better you're doing. A typical indicator lies between 46 and 50 percent.

Feed conversion measures the pounds (kilograms) of feed needed to produce a dozen eggs. This indicator is derived by dividing the total amount of feed eaten within a given period by the number of dozen eggs produced during that period. Commercial layer strains will always get better feed conversion than purebreds. For both, you can improve this indicator by reducing feed use (e.g., by controlling rodents that pilfer feed), increasing production (e.g., by culling lazy layers), or both. A good conversion rate is 4 pounds (1.8 kg) of feed or less per dozen eggs produced.

Feed efficiency, sometimes called "feed cost per dozen eggs," is derived by multiplying the feed conversion factor by the per-pound (kilogram) cost of feed. The lower your feed efficiency indicator, the better you're doing. On average, feed cost per dozen eggs represents approximately 65 percent of the total cost of egg production.

Eggs per hen per year measures the flock average in terms of dozens, allowing for losses and culls among the hens. A flock average of 20 dozen eggs per hen per year is pretty good. The loving keeper may coax out more eggs; the neglectful keeper will get fewer.

Depreciation of layers measures the second greatest cost of producing eggs — the difference between the value of hens at the beginning of the production period and their value at the end. Culls and deaths during the production period increase depreciation, as does advancing age. Younger hens may have some remaining value as layers; older hens are worth nothing more than cheap meat. The lower your flock's depreciation, the better you're doing.

Depreciation per dozen eggs is the depreciation of layers divided by the number of eggs laid during the production period. The lower your depreciation per dozen eggs, the better. Keep this indicator low both by keeping down depreciation of layers and by keeping up their production level. Unfortunately, sometimes the same management decision works both ways: culling poor layers improves production but increases depreciation.

Price of eggs to meet costs is the average total cost of maintaining each hen divided by the number of dozen eggs the average hen laid during the

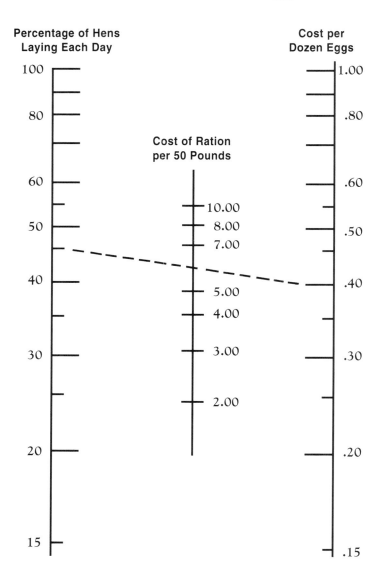

Feed Cost Per Dozen Eggs*

Percentage of Hens Laying Each Day

Cost per Dozen Eggs

100
80
60
50
40
30
20
15

Cost of Ration per 50 Pounds

10.00
8.00
7.00
5.00
4.00
3.00
2.00

1.00
.80
.60
.50
.40
.30
.20
.15

*Place a ruler across the percentage of hens laying per day and cost of ration to find the feed cost per dozen eggs.

Adapted from: "Farm Flock Management Guide," The Pennsylvania State University.

production period. This indicator represents your break-even point. If you keep layers strictly for family use, this indicator lets you compare the cost of producing your own eggs to the cost of purchasing eggs at the store. If you sell eggs, anything more you get per dozen must cover both your labor and profit on your layer-flock investment.

Net return per dozen eggs is the total amount earned from egg sales added to income from other sources (sale of spent hens, composted litter, etc.), less total expenses, divided by the number of dozen eggs produced during the production period. Net return is your bottom line.

CHAPTER 7

TABLE EGGS

FROM THE MOMENT AN EGG IS LAID, it begins to decline in quality. How slowly or rapidly this decline occurs depends on your methods of collection and storage. Once eggs are properly stored, a number of additional methods are available for prolonging their shelf-life for use in a variety of culinary dishes.

Egg Collection

Collect eggs often — preferably two or three times a day — so they won't get dirty or cracked, and so they won't start spoiling in warm weather or freeze in cold weather. Carry eggs in a small bucket or a basket in which they can't bang together or roll around.

If your nests are properly designed and managed, eggs should be clean when you collect them. A slightly dirty egg may be brushed off or rubbed with fine grain sandpaper. A seriously dirty egg may be rinsed in water that's slightly warmer than the egg (cooler water may force bacteria through the shell into the egg). Dry each egg before placing it in the carton. Avoid getting into the habit of routinely washing eggs, since water rinses off the natural bloom that helps preserve freshness.

Store eggs in clean cartons. An enclosed carton keeps eggs fresh longer than a carton that's opened at the top so you can see the eggs without lifting the lid. Orient eggs with their pointed end downward to keep their yolks nicely centered. Since an egg kept at room temperature ages the same amount in a day as a refrigerated egg ages in an entire week, refrigerate eggs as soon as possible if you plan to sell or eat them. (For information on the storage of hatching eggs, see page 195).

AMERICAN EGG BOARD

Store eggs in a clean carton, large ends up.

Egg Quality

Commercial eggs are sorted — according to exterior and interior quality — into four grades established by the USDA: AA, A, B, and Inedible. Nutritionally, there is no difference between one grade and another.

Grades AA and A are nearly identical, the main difference being that Grade A eggs are slightly older than Grade AA eggs. Grade B eggs have stained or abnormal shells, minor blood or meat spots, and other defects. They are used for baking and for making products such as powdered scrambled eggs. Inedible eggs are old, moldy, musty, or partially incubated and therefore unfit for human consumption.

Exterior Quality

Exterior quality refers to a shell's appearance, cleanliness, and strength. Appearance is important because the shell is the first thing you notice about an egg. Cleanliness is important because the shell is the egg's first defense against bacterial contamination; the cleaner the shell is, the easier it can do its job. Strength influences the egg's ability to remain intact until you're ready to use it.

The shell accounts for about 12 percent of the weight of a large egg and is made up of three layers:

- ◆ The inner, or mammillary, layer encloses the inner and outer membranes surrounding the egg. Between these two membranes is the air cell that develops at the large end as the egg ages.

◆ The spongy, or calcareous, layer is made up of tiny calcite crystals consisting of 94 percent calcium carbonate with small amounts of other minerals. Viewed through a microscope, these crystals look like thousands of thin pencils standing on end. The spaces between them form pores connecting the surfaces of the inner shell and outer shell so moisture and carbon dioxide can get out of the egg and air can get in to create the air cell. The pigment that gives the shell its color is laid over this middle layer.

◆ The bloom, or cuticle, is a light coating that seals the pores to preserve an egg's freshness by reducing evaporation and preventing bacteria from entering through the shell. Sometimes you'll find a freshly laid egg before the bloom has dried. When you wash an egg, the bloom dissolves, making the egg feel temporarily slippery. To replace natural bloom, commercial producers spray shells with a thin film of mineral oil, which is why store-bought eggs sometimes look shiny.

Except for preserving the freshness of eggs, shells have no culinary use (although I was once given a blended health-food drink containing a raw egg, shell and all, and I must admit it was pretty good). Shells have plenty of other uses. They can be:

◆ dried, crushed, and fed back to hens as a calcium supplement
◆ added to compost to sweeten the soil
◆ placed in tomato planting holes to prevent blossom-end rot
◆ decorated for a variety of arts and crafts

AMERICAN EGG BOARD

Shell Strength

A shell's strength is influenced by the vitamins and minerals in a hen's diet, especially vitamin D, calcium, phosphorus, and manganese. Shell strength is also influenced by a hen's age — older hens lay larger eggs with thinner, weaker shells.

Painted eggs make beautiful and lasting decorations.

A shell gets strength from its shape as well as from its composition. The curved surface is designed to distribute pressure evenly, provided the pressure is applied at the ends of the egg, not at the middle. The middle of a shell must be weak enough to allow a chick to peck all around and break out. Chefs take advantage of this characteristic to make a big show of breaking an egg with one hand — what you don't see is the thumb they press against the middle of the shell. By contrast, the ends of an egg must be quite strong so a newly laid egg won't break when it lands in the nest.

Use a bathroom scale to measure egg-shell strength.

One way to test the strength of an egg is to press the ends between the palms of your hands. For a more precise measurement, use an ordinary bathroom scale. Stack some boards or books to equal the height of the egg standing on end. With paper towels, fashion a ring around the bottom of the egg to stand it on end, next to the books. Rest one edge of the scale on the books and the other edge on the egg. Press on the scale just above the egg. A well-formed shell should support up to 9 pounds (4 kg) before it breaks.

Interior Quality

Interior quality refers to the appearance and consistency of an egg's contents. The quality of both the white and the yolk deteriorate as an egg ages. Quality may have been poor to start with, depending on the health of the hen, the use of medications, the weather, and hereditary factors.

An egg's interior qualities can easily be determined by breaking the egg into a dish for examination. Experts learn to judge the quality of intact eggs by candling (as described on page 211). If you have never candled an egg, practice with white-shelled eggs before you tackle brown ones, which are more difficult to see through.

Grasp an egg between your thumb and first two fingers. With the egg at a slant, hold the large end to the candling light. Turn your wrist to give the egg a quick twist, sending the contents spinning.

The white, or *albumen*, of a fresh egg is fairly dense. As the egg ages, the white grows thinner, letting the yolk move more freely. When you twirl the egg during candling, the deteriorating albumen lets the yolk move closer to the shell. The older the egg, the more easily you can see its yolk.

An egg has more than one kind of albumen, as you will discover if you break an egg and examine its contents. Two clearly visible kinds are the firm white around the yolk and the thin white closer to the shell. The outer thin white, as it is called, repels bacteria by virtue of its alkalinity and its lack of the nutrients needed by bacteria for growth. The firm or thick white surrounding the yolk cushions the yolk and provides additional chemical defenses against bacteria. A less obvious second thin layer lies between the firm white and the yolk. The older an egg gets, the more thin white and the less thick white it has.

Another kind of white is the *chalaziferous layer*, made up of dense albumen surrounding the yolk. The ends of this layer are twisted together to form a *chalaza* (pronounced kah-lay'-za) on either side of the yolk. These two cords anchor the chalaziferous layer and protect the yolk by centering it within the white. During its formation, as the egg travels through the oviduct and rotates, the chalazae become twisted. When you break an egg into a dish, the chalazae snap away from the shell and recoil against the yolk. Misinformed cooks mistake the resulting two white blobs at opposite sides of the yolk for the beginnings of a developing chick.

The albumen of a fresh egg contains carbon dioxide that makes the white look cloudy. As an egg ages, the gas escapes and the white turns more transparent. A yellowish or greenish hue to the albumen of a fresh egg indicates the presence of the B vitamin riboflavin.

Anatomy of an Egg

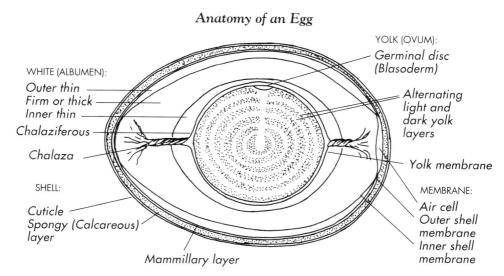

YOLK (OVUM):
Germinal disc (Blasoderm)

WHITE (ALBUMEN):
Outer thin
Firm or thick
Inner thin
Chalaziferous
Chalaza

Alternating light and dark yolk layers

Yolk membrane

SHELL:
Cuticle
Spongy (Calcareous) layer

MEMBRANE:
Air cell
Outer shell membrane
Inner shell membrane

Mammillary layer

Yolk Color

Egg yolks get their color from *xanthophyll,* a natural yellow-orange pigment in green plants and yellow corn, and the same pigment that colors the skin and shanks of yellow-skinned hens. The exact color of a yolk depends on the source of the xanthophyll. Alfalfa, for example, produces a yellowish yolk while corn gives yolks a reddish-orange color.

Excessive amounts of certain pigmented feeds can affect yolk color. Alfalfa meal, kale, rape, rye pasture, and certain weeds including shepherd's purse, mustard, and pennycress turn yolks dark. Too much cottonseed meal can really throw off yolk color, causing it to be salmon, dark green, or nearly black.

Yolk Color

Color	Cause
RAW YOLKS	
reddish, olive green, black	green grass, silage, cottonseed meal
green	acorns, shepherd's purse
orange to dark yellow	green feed, yellow corn
dark yellow	marigold petals
medium yellow	alfalfa meal, yellow corn
pale yellow	white corn
	coccidiosis (rare)
COOKED YOLKS	
yellow rings	normal layers of yolk
gray or green surface	overcooked yolk
	iron in cooking water
green rings	iron in hen's feed or water
greenish when scrambled	overcooked egg and/or egg
	held too long before serving
greenish when soft-cooked	served with blueberry pancakes

Fresh Egg Tests

Unless your hens are caged, you will occasionally find an egg, or a cache of eggs, whose age may be in question. Five different methods allow you to estimate an egg's age.

Yolk visibility is determined by giving the egg a sharp twist in front of a candling light and viewing the spinning contents. If the yolk looks vague and fuzzy, the thick white surrounding it is holding it properly centered within the shell. A yolk that's clearly visible has moved closer to the shell, indicating albumen that has thinned with age.

Air-cell size increases as an egg ages. A freshly laid egg has no air cell. As the egg cools its contents shrink. The inner shell membrane pulls away from the outer shell membrane, forming a cell, or pocket, at the large end of the egg. As moisture evaporates from the egg, its contents continue to shrink and the air cell grows. Candling to measure the cell will give you an idea of the egg's age. The cell of a freshly laid cool egg is no more than ⅛-inch (3 mm) deep. From then on, the larger the cell, the older the egg. Just how fast the air cell grows depends on the porosity of the shell and on the egg's storage temperature and humidity.

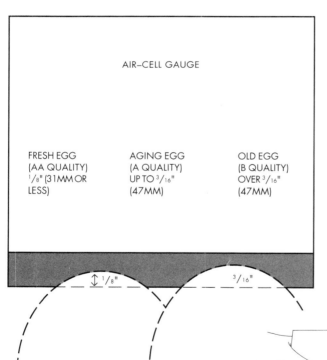

AIR–CELL GAUGE

FRESH EGG
(AA QUALITY)
¹/₈" (31MM OR
LESS)

AGING EGG
(A QUALITY)
UP TO ³/₁₆"
(47MM)

OLD EGG
(B QUALITY)
OVER ³/₁₆"
(47MM)

↕ ¹/₈"

³/₁₆"

Photocopy gauge, paste over cardboard, and cut out (or order an "Official Egg Air Cell Gauge" Form PY-35 from Agricultural Marketing Service, USDA, Washington, DC 20250). To measure an air cell, place gauge over the large end of an egg held in front of a light.

Float Test

An easy way to determine an egg's age is by placing it in water. Older eggs have larger air cells and will therefore float.

Floating an egg in plain water lets you gauge its air-cell size without candling. A fresh egg will settle to the bottom of the container and rest horizontally. The larger air cell of a 1-week-old egg will cause the big end of the egg to rise up slightly from the container bottom. An egg that's 2–3 weeks old will settle to the bottom of the container vertically, big end upward. The large air cell of a very old egg causes the egg to float.

Smell is the quickest way to detect the age of a very old egg. A rotting egg emits foul-smelling hydrogen sulfide, otherwise known as "rotten-egg gas."

Breaking an egg and examining its contents is another way to estimate the egg's age. A fresh egg has cloudy, firm albumen that holds the yolk up high; a stale egg has clear, watery albumen that spreads out thinly around the yolk. The older the egg is, the greater the likelihood that its yolk will break. As an egg ages, water migrates from the albumen to the yolk, stretching and weakening the yolk membrane. This test isn't foolproof, as even the freshest egg occasionally has a watery white or an easily broken yolk.

Egg Abnormalities

You may on occasion find an egg that is abnormal due to an accidental occurrence, a hen's hereditary tendencies, or environmental or management factors. Some egg abnormalities are little more than nonrecurring curiosities. Others may require corrective action on your part. You can detect these abnormalities by candling and by inspecting broken-out eggs. Trapnesting (as described on page 174) lets you identify and cull hens that habitually lay problem eggs.

No-yolkers are called "dwarf" or "wind" eggs. Such an egg is most often a pullet's first effort, produced before her laying mechanism is fully geared up. In a mature hen, a wind egg is unlikely, but can occur if a bit of reproductive tissue breaks away, stimulating the egg-producing glands to treat it like a yolk and wrap it in albumen, membranes, and a shell as it travels through the egg tube. You can tell this has occurred if, instead of a yolk, the egg contains a small particle of grayish tissue. In the old days, no-yolkers were called "cock" eggs. Since they contain no yolk and therefore can't hatch, our forebears believed they were laid by roosters.

Double yolkers appear when ovulation occurs too rapidly, or when one yolk somehow gets "lost" and is joined by the next yolk. Double yolkers may be laid by a pullet whose production cycle is not yet well synchronized. They're occasionally laid by heavy-breed hens, often as an inherited trait.

Occasionally an egg contains more than two yolks. I once found a pullet's egg that had three. The greatest number of yolks ever found in one egg is nine. Record-breaking eggs are likely to be multiple yolkers: The *Guinness Book of Records* lists the world's largest egg (with a diameter of 9 inches/22.5 cm) as having five yolks and the heaviest egg (1 pound/0.45 kg) as having a double yolk and a double shell.

Egg within an egg, or a double-shelled egg, appears when an egg that is nearly ready to be laid reverses direction and gets a new layer of albumen covered by a second shell. Sometimes the reversed egg joins up with the next egg, and the two are encased together within a new shell. Double-shelled eggs are so rare that no one knows exactly why or how they happen.

Blood spots occur when blood or a bit of tissue is released along with a yolk. Each developing yolk in a hen's ovary is enclosed inside a sac containing blood vessels that supply yolk-building substances. When the yolk is mature, it is normally released from the only area of the yolk sac, called the "stigma" or "suture line," that is free of blood vessels. Occasionally the yolk sac ruptures at some other point, causing vessels to break and blood to appear on the yolk or in the white. As an egg ages, the blood spot becomes paler, so a bright blood spot is a sign that the egg is fresh.

Blood spots occur in less than one percent of all eggs laid. They may appear in a pullet's first few eggs, but are more likely to occur as hens get older, indicating that it's time to cull. Blood spots may be triggered by too little vitamin A in a hen's diet, or they may be hereditary — if you hatch replacement pullets from a hen that characteristically lays spotty eggs, your new flock will likely do the same.

Meat spots are even less common than blood spots. They appear as brown, reddish brown, tan, gray, or white spots in an egg, usually on or near the yolk. Such a spot may have started out as a blood spot that changed color due to chemical reaction, or it may be a bit of reproductive tissue. Since meat spots look unappetizing, cull a hen whose eggs characteristically contain them.

Developing embryos are a sign that eggs have been partially incubated, which occurs when hens can steal, or hide, their nests so that eggs are allowed to accumulate. If you find a bunch of eggs and you can't be sure how long they've been there, inspect them either by candling or by breaking them into a separate dish before using them. Developing embryos will not be a problem if your henhouse is set up so that eggs can't be hidden from you or if you have no rooster to fertilize eggs.

Wormy eggs are extremely rare, occurring only in hens with a high parasite load. Finding a worm in an egg is not only unappetizing but is a clear indication that you are not doing a good job in keeping your hens healthy and parasite-free.

Off flavor may result from something a hen ate or from environmental odors. Hens that eat onions, garlic, fruit peelings, fish meal, and fish oil will lay eggs with an undesirable flavor. Eggs can also absorb odors that translate into unpleasant flavors if they're stored near kerosene, carbolic acid, mold, must, fruits, and vegetables.

Egg Safety

A freshly laid egg is warm and moist, and therefore attractive to a variety of bacteria and molds in the environment. As the egg cools and its contents shrink, these microbes may be drawn through the shell's 6,000-plus pores. Your first line of defense in keeping eggs safe to eat is to keep nests clean and lined with fresh litter. Eggs produced in a clean environment, collected often, and promptly refrigerated contain too few microbes to cause human illness.

An egg that's cracked — called a "check" egg — is safe to eat, provided the membrane is intact, the egg is refrigerated promptly, and it's used within a week. If a cracked egg leaks, indicating that the membrane has been broken through, discard it.

Discard eggs that are seriously soiled. Although moderately soiled eggs may be washed, in doing so you'll rinse away the bloom that seals the pores and keeps out bacteria. In that event, sanitize the cleaned shell by dipping the egg in a solution of water and chlorine bleach. Dry the egg well before placing it in a carton.

Clean eggs stored at 45°F (7°C) and 70 percent humidity will keep well for at least three months. In a standard household refrigerator, where foods tend to dry out, eggs will keep for up to five weeks.

Certain species of bacteria (such as *Streptococcus* and *Staphylococcus*) tolerate dry conditions and are therefore able to survive on an egg's shell. These bacteria dwindle during storage, but are replaced by other species (including *Pseudomonas*) that cause eggs to rot. The presence of *Pseudomonas* can be detected by a deep-green color in an egg's white and perhaps by a sour smell emitted when the egg is broken open. Molds (*Penicillium, Alternaria, Rhizopus*) that get on the shells of eggs washed in dirty water or stored in humid conditions may also penetrate the shell and cause spoilage.

Salmonella may be either on the shell or inside the egg when it is laid. These bacteria can cause serious illness in humans but only if allowed to multiply, which happens when a contaminated egg is held for too long at room temperature. The problem gets worse if the egg is combined with other eggs in a mixture that's left on the counter and then is undercooked or served as is.

Cooking eggs destroys *Salmonella* bacteria. A thoroughly cooked egg has its white cooked through and its yolk at least beginning to thicken — the yolk need not be hard, but should no longer be runny. Always cook an egg slowly to make sure it's heated all the way through.

Keep hot foods containing eggs at 140°F (60°C) or warmer, and cold foods at 40°F (4°C) or cooler. Acid ingredients, including lemon juice or vinegar, in foods containing raw eggs retard bacterial growth. Even so, such foods should be eaten immediately or refrigerated promptly.

To avoid food poisoning from eggs:

- Collect eggs often and refrigerate them promptly.
- Discard seriously cracked or dirty eggs.
- Wash hands and utensils after handling raw eggs.
- Immediately cook or refrigerate foods prepared with raw or undercooked eggs.
- Cook eggs and egg-rich foods to 160°F (71°C).
- Promptly refrigerate leftovers.

Nutritional Value

Despite all the bad press eggs get, they are still considered the "perfect" food. One egg contains almost all the nutrients necessary for life, lacking only vitamin C. A large egg weighing 2 ounces (46 g) is approximately 31 percent yolk, 58 percent white, and 11 percent shell.

The yolk is basically an oil-water emulsion containing proteins, fats, cholesterol, pigment, and a number of minor nutrients. Yolks contain lecithin, which acts as an emulsifier for making mayonnaise and hollandaise sauce, and also affects human brain function. Unfortunately, most of an egg's cholesterol and calories are in the yolk. The caloric value of an egg depends on its size.

Caloric Content of Eggs

Size	Calories
Peewee	47
Small	56
Medium	66
Large	75
Extra large	84
Jumbo	94

The white is made up of water combined with several different kinds of protein. Egg protein is complete, since it contains all the essential amino acids. It is among the highest quality protein found in food, second only to mother's milk. One large egg contains approximately 6.25 grams of protein, or about 15 percent of the U.S. recommended daily allowance (RDA). A large egg is roughly equivalent in protein to 1 ounce (28 g) of lean meat, poultry, fish, or legumes. The United Nations Food and Agriculture Organization (FAO) rates the biological values of the top five protein sources, on a scale where 100 is the most efficient, as follows:

whole egg	93.7
cow milk	84.5
fish	76.0
beef	74.3
soybeans	72.8

One of the proteins in raw eggs, *avidin*, ties up the vitamin biotin as one of an egg's defenses against bacteria, since most bacteria can't grow without biotin. Pets are sensitive to the effects of avidin and shouldn't be routinely fed raw eggs. A human would have to eat two dozen raw eggs a day to be affected. Avidin is inactivated by heat.

Many people have the mistaken idea that fertile eggs are more nutritious than infertile eggs. The idea is encouraged by unscrupulous sellers who cater to the health-food market and who feel they can charge more by claiming fertile eggs are more nutritious than the infertile eggs commonly sold in supermarkets. In truth, a sperm contributes an insignificant amount of nutrients to a fertilized egg.

Another idea promoted in health-food circles is that eggs with colored shells are more nutritious than white-shelled eggs. Although eggs from backyard or range-fed flocks are likely to have brown or blue-green shells, in contrast to the white shells of commercial eggs, shell color per se has nothing to do with an egg's nutritional content.

Nutrition Facts

Serving Size 1 egg (50g)
Serving per Container 12

Amount Per Serving

Calories 70 Calories from Fat 40

	% Daily Value*
Total Fat 4.5g	**7%**
Saturated Fat 1.5g	**8%**
Polyunsaturated Fat .5g	
Monounsaturated Fat 2.0g	
Cholesterol 215mg	**71%**
Sodium 65mg	**3%**
Potassium 60mg	**2%**
Total Carbohydrate 1g	**0%**
Protein 6g	**10%**

Vitamin A 6% · Vitamin C 0%
Calcium 2% · Iron 4% · Thiamin 2%
Riboflavin 15% · Vitamin B-6 4%
Folate 6% · Vitamin B-12 8%
Phosphorus 8% · Zinc 4%

Not a significant source of Dietary Fiber or Sugars.

*Percent Daily Values are based on a 2,000 calorie diet. Your daily values may be higher or lower depending on your calorie needs.

		Calories	2,000	2,500
Total Fat	Less than		65g	80g
Sat Fat	Less than		20g	25g
Cholesterol	Less than		300mg	300mg
Sodium	Less than		2,400mg	2,400mg
Potassium			3,500mg	3,500mg
Total Carbohydrate			300g	375g
Dietary Fiber			25g	30g
Protein			50g	65g

Calories per gram
Fat 9 · Carbohydrate 4 · Protein 4

Cholesterol

Cholesterol is a kind of fat found in the bodies of chickens and humans alike. It is required for both the synthesis of vitamin D from sunshine and the production of sex hormones. But it can also collect in the bloodstream and clog the arteries. Thanks to media scare tactics, many people have the impression that all the cholesterol you eat goes straight into your bloodstream. Not true.

About 80 percent of the cholesterol in your body is manufactured within your body. The amount of manufactured cholesterol is regulated by your liver, based on your dietary intake. Some people have better controlling mechanisms than others.

Only about 20 percent of your blood cholesterol comes directly from the food you eat. Saturated fat increases blood cholesterol, monounsaturated fat has no apparent effect, polyunsaturated fat tends to decrease blood cholesterol. Dietary cholesterol somewhat increases blood cholesterol, but a healthy body compensates by producing less and/or excreting more to maintain the proper blood cholesterol level. The accompanying "Nutritional Facts" label lists the amounts of fat and cholesterol in the average large-size egg.

Several studies have shown that eating no eggs, compared to eating a moderate number, does not affect blood cholesterol. Consider the 88-year-old man who, as reported in the *New England Journal of Medicine*, ate 25 eggs a day for more than 15 years, yet had normal blood cholesterol and no sign of heart disease. Consider the group of healthy young men in another study who ate as many as 14 eggs a week without any significant change in their blood cholesterol. Consider me: I eat an egg each morning for breakfast and sometimes have a couple more for lunch or dinner. Since acquaintances keep asking about my cholesterol level, I had it checked. To quote my doctor, it's "perfect."

Whether or not eggs are harmful seems to depend on who's eating them. An estimated two-thirds to three-quarters of the population can handle a moderate number of eggs. Medical practitioners who don't buy into the cholesterol panic point out that, assuming *all* the cholesterol in an egg went into your bloodstream, you'd have to eat five jumbo eggs a day to raise your cholesterol level from 150 mgm% to 152 mgm%. Perhaps one reason eggs do less harm than some researchers predict is that they contain lecithin, which nutritional experts believe scours cholesterol from the arteries.

Cholesterol Reduction

I'm often asked if some kinds of eggs contain less cholesterol than others, especially since sellers sometimes charge outrageous prices for blue-green eggs, claiming they're lower in cholesterol than white-shelled eggs. Studies show that eggs laid by hens of heavier breeds (including Araucana) are likely to contain slightly *more* cholesterol than eggs laid by commercial Leghorn strains. In explaining these findings, researchers reason that the more often a hen lays, the less time she has to put cholesterol into each egg. Unfortunately, the difference among breeds is so slight as to be insignificant. Otherwise, commercial producers wouldn't have such a hard time trying to reduce the cholesterol level of eggs through genetic selection.

Other sellers claim that fertile eggs and those laid by range-fed hens have less cholesterol than supermarket eggs. Fertilization, per se, does not significantly affect an egg's nutritional value. Range-feeding *may* affect an egg's level of saturated fat, although the jury is still out. Preliminary studies (reported by Joel Salatin and others) indicate that the orange yolks and yellow fat in range-fed hens reflect low levels of saturated fat.

The only other property known to influence fat is an egg's size — the larger the egg, the more white it has in proportion to yolk, and therefore the lower its percentage of saturated fat and cholesterol. Most of the fat and all of the cholesterol are in the yolk. For the record, cooking an egg does not affect fat or cholesterol.

According to some nutritionists, you can eliminate more saturated fat from your diet by eating chicken without the skin than by eliminating eggs from your diet. Still, if you wish to reduce fat in an egg dish (such as a quiche or an omelette), substitute 2 egg whites per whole egg for half the eggs called for in the recipe. To reduce saturated fat in baked goods, substitute 2 egg whites and 1 teaspoon of vegetable oil for each whole egg. If the recipe already has oil in it, leave out the extra oil.

So-called egg "substitutes" are made from egg white. The yolk is replaced by such ingredients as nonfat milk or tofu, along with a variety of emulsifiers, stabilizers, antioxidants, gums, and artificial coloring. This list makes you wonder if the healthier choice wouldn't be to eat the real thing.

The Healing Egg

In folk medicine, egg whites are used to heal wounds and inhibit infection. Egg white contains *conalbumin,* a substance that binds iron so it can't be used by microbes. By inhibiting bacterial growth in an egg, conalbumin protects a developing embryo from infections. To treat a wound and speed healing, the protein-rich membrane inside the shell is peeled away and bandaged in place over a cut. Raw eggs are also used as beauty aids — whites in facials, yolks in shampoos and hair conditioners.

Egg Cuisine

If you pride yourself on your culinary talents with eggs, try your hand in an egg-cooking contest. For entry dates and other regulations pertaining to local contests, contact either your state Egg and Poultry Association or your state Department of Agriculture. Winners in local cook-offs vie for big money in the National Egg Cooking Contest sponsored by the American Egg Board (as listed in the appendix).

Perhaps the greatest tribute to eggs is the vast number of ways they're prepared by cooks around the world. Many recipes call for eggs as a functional ingredient, used to perform such services as binding other ingredients together, clarifying broth, or causing baked goods to rise. Many additional recipes highlight the egg itself. One of the most versatile of these is the hard-cooked egg.

Function of Eggs as an Ingredient

Function	Example
Bind	meatloaf
Clarify	soup
Emulsify	mayonnaise
Garnish	salad
Glaze	cookies
Leaven	sponge cake
Retard crystallization	frosting
Thicken	custard

From: *National 4-H Avian Bowl Manual.*

Hard-Cooked Eggs

Hard-boiled eggs have many uses, ranging from picnic garnish to afternoon snack, from main dish at lunch to appetizer before dinner. Actually, the term "hard-boiled" is inaccurate. True, the egg is cooked hard, but if it has been *boiled* it will be rubbery and tough. A more accurate term is "hard-cooked." The old joke about a novice chef not being able to so much as boil an egg isn't far-fetched — there are several tricks to properly preparing a hard-cooked egg.

For starters, if you've never tried to peel a hard-cooked fresh egg, you're in for a shock. The contents of a fresh egg fill up nearly the entire shell and, when the egg is cooked through, stick to the shell. If you try to peel the egg, you'll peel off several layers of white, too. The contents of an egg that's at least a week old have shrunk away from the shell, making it easier to peel. Store-bought eggs peel so easily because they're usually at least a week old by the time you get them home.

Old eggs have an off-center yolk, which doesn't look as nice as a centered yolk when the cooked egg is sliced or deviled. The yolk becomes displaced from the center as the egg ages and its white thins. The fresher the egg, the more centered the yolk, but the more difficult the egg is to peel. A good compromise is to hard-cook eggs that are 7 to 10 days old. If you wish to hard-cook eggs that are fresher, leave them overnight at room temperature first.

Another way to make a fresh egg easier to peel is to poke a hole in the large end with a pin or tack. Place eggs in a single layer in a saucepan. Add water to cover the eggs by at least 1 inch (2.5 cm) and bring the eggs to a boil. Put on the lid, remove the pan from the heat, and let it sit 15 minutes for small eggs, 18 minutes for medium eggs. Remove the eggs from the hot water with a slotted spoon and place them in ice water for 1 minute while you bring the water in the saucepan back to a simmer. One by one remove the eggs from the cold water and place them in the simmering water for exactly 10 seconds, just long enough for the shell to expand away from the rest of the egg. Peel the egg immediately by cracking the shell all over and rolling the egg between your palms. Start peeling at the large end, holding the egg under running water to rinse away bits of shell.

Hard-cooked eggs have many uses, from garnish to featured dish.

To prevent cracking, cover cold eggs with cold water, then bring the water to a boil.

Another problem is cracked shells that occur when cold eggs are dropped into boiling water. To prevent cracked shells, place cold eggs in an empty saucepan, cover them with cold water, then bring the water to a boil.

A third problem is the unappetizing greenish tinge that develops around the edges of a cooked yolk. As soon as the eggs have been in the hot water long enough to cook through, cool them quickly and the yolks won't turn green. Cool the eggs either by putting them in ice water or by running cold tap water over them for 5 minutes.

To avoid confusing raw eggs with cooked eggs, pencil an "X" on the latter. Refrigerate the eggs as soon as they have cooled. When you first put them into the refrigerator, the eggs will emit a gassy odor caused by hydrogen sulfide that formed while they were cooking. After a few hours the odor will go away by itself. Store hard-cooked eggs no longer than two weeks.

Salted Eggs

In some circles, salted eggs are a delicacy. There are basically two methods for preparing salted eggs: salt them and then cook them, or cook them and then salt them.

To salt eggs before cooking them, boil 3 cups of water with 6 cups of salt, and cool the brine solution. In a wide-mouth 1-quart jar, place 6 raw unshelled eggs. Pour the undissolved portion of the salt over the eggs and top it off with the supersaturated brine solution. Weigh down the eggs with a water-filled well-sealed plastic bag. Cover the jar with perforated paper and store it in a cool place for 12 days. To cook the salted eggs, cover them with fresh water, bring the water to a boil, and simmer for 5 minutes. Turn off the heat and let the eggs stand in the hot water for 10 minutes more. Cool, peel, and serve.

To salt eggs after they have been hard cooked (called *soleier*), bring 3 cups of water to a boil with ¼ cup salt (for a little color, add the skins of three yellow onions). Simmer the mixture for 10 minutes or until the water turns brown. Remove the solution from the heat and let it cool. Crack the shells of 6 hard-cooked eggs all around, but do not peel them. Put the eggs into a wide-mouth 1-quart jar and pour the solution over them (straining out the onion skins).

Refrigerate the eggs for at least 24 hours before serving them. Slice or halve the eggs and serve them with mustard.

Beating Egg Whites

A gourmet cook beats egg whites only in a copper bowl. Copper from the bowl reacts with conalbumin in the white to stabilize the foam so the air that's beaten in will not leak right back out. You can get the same results in a glass or stainless steel bowl by adding an acid ingredient.

Cream of tartar, found in the spice section at grocery stores, is an acid ingredient. Add ⅛ teaspoon per egg white unless you're making meringue, in which case add ⅛ teaspoon per two egg whites. Lemon juice or vinegar serves the same purpose as cream of tartar.

For a stable meringue, add cream of tartar or other acid ingredient.

Fat-free meringue cookies may be made in many tasty flavors.

Fat-Free Meringue Cookies

2 large egg whites	⅝ cup sugar
⅛ teaspoon salt	½ teaspoon vanilla
¼ teaspoon cream of tartar	⅛ teaspoon almond extract

1. Beat egg whites until foamy.

2. Add the salt and cream of tartar. Beat to soft peaks.

3. Gradually beat in the sugar. When the sugar has dissolved, add the vanilla and almond extract. (For variety, add ¼ cup chopped nuts, candied fruit, toasted coconut, or shaved chocolate to the batter before forming cookies.)

4. On a greased cookie sheet, closely space 48 mounds of meringue containing one tablespoon each. Bake for 1 hour at 225°F (107°C). Let the cookies cool in the oven for 1 hour before removing them from the sheet.

Egg whites will beat up to their greatest volume if the eggs are first warmed to room temperature for about 30 minutes. Since fat inhibits beating, take great care not to get any yolk into the white when you separate the eggs. Properly beaten whites should increase to four times their volume. If you overbeat the whites to the point where they turn stiff and dry, restore them by beating in an additional room-temperature white.

Leftover Yolks

If you use a lot of egg whites and you have no immediate need for the leftover yolks, hard cook them to use as a garnish on salads and other dishes. Place the yolks in a single layer in a saucepan, add enough water to cover them by at least 1 inch. Place the cover on the saucepan and bring the water to a boil. Remove the pan from the heat and let it stand, covered, for 15 minutes. With a slotted spoon, remove the yolks from the water and drain them well. Store the yolks in a sealed container in the refrigerator for no more than 2 days, or freeze them for later use. A delightful cookbook — now out of print, but well worth seeking in used bookstores — is *The Other Half of the Egg,* a collection of recipes for using separated whites and yolks.

Microwave Cooking

Microwaving is a fast way to prepare eggs without added oil. Use full power for scrambled eggs and omelets, 30 to 50 percent power for unbeaten eggs.

Since the fat in yolks causes them to cook more quickly than whites, steam can build up within the yolk membrane and cause a yolk to explode. Avoid a

Size Equivalents

Most recipes call for large eggs. If your hens lay eggs of some other size, use the following chart to figure out how many you need:

Large	Jumbo	Extra-Large	Medium	Small	Peewee
1	1	1	1	1	2
2	2	2	2	3	3
3	2	3	3	4	5
4	3	4	5	5	6
5	4	4	6	7	8
6	5	5	7	8	10

mess by pricking yolks with a toothpick or the tip of a knife to let the steam escape. Confine potential explosions and ensure more even cooking by covering eggs with microwaveable plastic wrap. Do not microwave whole eggs in their shells; they will definitely explode.

Preserving Eggs

At some times of the year you'll have more eggs than you can use, while at other times you'll have too few. It becomes logical to preserve surplus eggs in times of plenty to use in times of need.

Throughout the ages, different means have been devised to prolong the storage life of eggs. The ancient Chinese stored them for years in various materials ranging from clay to wood ashes to cooked rice. You probably wouldn't want to eat one of those eggs, with its greenish yolk and gelatinous brown albumen.

In modern times, many options have been developed for preserving eggs. Although none is an outright substitute for cold storage, these methods can let you extend shelf life. They also offer short-term ways to prolong storage without electricity, which can be handy if you live in or are planning a trip to the

Storage Methods and Maximum Storage Times

Method	Storage Time
Refrigeration	
whole in carton	5 weeks
whites, in tightly sealed jar	4 days
yolks, covered with water	2 days
hard cooked in shell	2 weeks
hard cooked, peeled, in water	1 week
hard cooked, peeled & pickled	6 months
Water glass	6 months at 34°F (1°C)
Oiled	7 months at 31°F (-0.5°C)
Thermostabilized	8 months at 34°F (1°C)
Oiled & Thermostabilized	8 months at 34°F (1°C)
Cold storage	9 months at 30°F (-1°C)
Frozen	12 months at 0°F (-18°C)

These guidelines are for fresh, home-produced eggs only.

outback. Even in the back woods, you can often take advantage of a cold running stream or an ice bank to keep eggs fresh. Many cellars offer suitable egg storage conditions. The cooler the temperature, the longer eggs will keep without spoiling.

Eggs can be safely stored for the short term (two to three months) at temperatures up to 55°F (13°C), where the relative humidity is close to 75 percent. The moisture level is important since at low humidities eggs dry out and at high humidities they get moldy. If the storage area is damp, mold on the shells needs to be wiped off before the eggs are used. A little air circulation helps retard mold growth.

At 30°F (–1°C) eggs will keep for as long as nine months. The temperature must not get below 28°F (–2°C), though, or the eggs will freeze and burst their shells. Relative humidity of 85 percent is best for long-term storage. To prevent mold growth and condensation, seal egg cartons in plastic bags. To minimize drying out at lower humidities, use thermostabilization, oil, or water glass, as described later in this chapter.

Preserve only eggs with clean uncracked shells. Dirty eggs that have been cleaned by washing or dry buffing do not keep well under prolonged storage. Given the infinite number of combinations of possible temperature and humidity conditions, it's not possible to list definitive storage times under all conditions. The accompanying guidelines, in combination with the fresh-egg tests offered earlier in this chapter, offer a good starting place. The longer you store eggs, the more likely they are to develop a stale or off flavor that makes them less suited for breakfast than for use in recipes.

Refrigeration

Refrigeration is often the quickest, most convenient way to store eggs. On the lowest shelf, where the temperature is coolest, eggs in a closed carton will keep for up to five weeks. The biggest problem with a household refrigerator is its low humidity, especially in a self-defrosting (frost-free) model. If you wrap egg cartons in plastic bags to prevent moisture loss (as well as absorption of flavors from other foods), you can refrigerate eggs for two months. Eggs on an egg rack on the door won't last long due to jostling, blasts of warm air whenever the door is opened, and lost moisture.

If you use only whites or yolks for a recipe, you can refrigerate leftover whites in a tightly covered container for up to four days. Cover leftover yolks with cold water and use them within two days.

Freezing

Freezing lets you keep eggs longer than any other method — up to one year at 0°F (–18°C). Freeze only raw eggs; hard-cooked eggs will turn rubbery. Since eggs that are frozen intact will expand and burst their shells, the shells must be removed. You may store eggs either whole or with the whites and yolks separated. To keep yolks or whole eggs from getting gummy, add sugar or salt, depending on how you plan to use them. Freeze the eggs in ice cube trays or in freezer containers.

Eggs that have been frozen in ice cube trays should be removed and wrapped in heavy freezer paper or plastic bags. To compare the size of a frozen cube with the normal size of an egg (or its yolk or white), use a measuring spoon to determine how much each slot in the tray holds, then consult the accompanying "Frozen Egg Equivalents" chart. If you use metal trays with removable grids, measure the total amount the tray holds and divide by the number of cubes per tray.

For use in recipes calling for several eggs, freeze eggs (or yolks or whites) in air-tight freezer containers, each holding just enough for one recipe. Leave a little head space to allow for expansion, otherwise the lid may pop or the container may split. Place a square of freezer paper on top of the eggs to minimize the formation of ice crystals. Label each container with the date, contents, and recipe for which it is intended.

To store whole eggs, break the contents into a bowl, stir just enough to blend the yolks with the whites (taking care not to whip in air), and press the eggs through a sieve to break up the thick albumen. You can get by without sugar or salt if the whites and yolks are thoroughly mixed. Otherwise, to each cup of eggs add ½ teaspoon salt (for a main dish) or ½ tablespoon honey, corn syrup, or sugar (for a dessert). Pour the mixture into trays or containers for freezing.

If you prefer, carefully separate the whites from the yolks, taking care to avoid getting any yolk into the whites, so the whites can be whipped later. Press the whites through a sieve to break up the thick albumen, and freeze them in

---

Frozen Egg Equivalents

2 T thawed white = 1 large fresh white
1 T thawed yolk = 1 large fresh yolk
3 T thawed whole egg = 1 large fresh egg

Egg Storage Safety

Temperature	°F	°C	Humidity
	100	38	
Quality is rapidly lost at these temperatures, the higher the temperature the faster the loss.	95	35	
	90	32	
	85	29	Molds appear and grow at relative humidities above wet bulb 52°F (11°C) or 85 percent relative humidity.
	80	26.5	
	75	24	
	70	21	
	65	18	
Short Storage Periods	60	15.5	
	55	13	
Long-term Storage Quality is retained best at these temperatures.	50	10	**Safe:** Wet bulb 42°F to 52°F (5.5°C to 11°C) or 75 to 85 percent relative humidity.
	45	7	
	40	4.5	
	35	1.5	
	30	-1	
	25	-4	
	20	-6.5	Eggs lose quality as a result of moisture loss when relative humidity is below wet bulb 42°F (5.5°C) or 75 percent.
Eggs break at temperatures below 28°F (–2°C).	15	-9.5	
	10	-12	
	5	-15	
	0	-18	

Adapted from: "Table Egg Room Safety Chart," University of Georgia.

trays or containers. Thawed whites can be whipped just like fresh ones if you let them warm to room temperature for 30 minutes before beating them.

Separated yolks need sugar or salt to prevent gumminess. Add either ⅛ teaspoon salt or ½ tablespoon honey, corn syrup, or sugar per 4 yolks (approximately ¼ cup or 60 ml). Freeze the yolks in ice cube trays or air-tight containers.

Thaw only as many frozen eggs as you can use within three days. Thaw them overnight in the refrigerator or in air-tight containers placed in cool water, 50°F to 60°F (10–16°C). Use thawed eggs only in foods that will be thoroughly cooked.

Pickling

Pickling is a good way to preserve hard-cooked eggs. Pickled eggs may be used in place of hard-cooked eggs in salads or in place of pickles in sandwiches. They also make a great snack.

Over the years, I have prepared many dozens of pickled bantam eggs, packed hot in boiling vinegar and processed in sealed pint jars in boiling water

Old-Fashioned Pickled Eggs

    4–6   whole cloves
       1   stick cinnamon
       2   cups white vinegar
       ½   teaspoon dry mustard
       ½   teaspoon salt
       ½   teaspoon pepper
       6   hard-cooked eggs

1. Place the cloves and cinnamon in a saucepan with the vinegar, and bring mixture to a boil.
2. Blend the mustard, salt, and pepper with a little water and stir into boiling vinegar. Simmer for 5 minutes.
3. Peel the eggs and place them in a jar.
4. Pour hot mixture over eggs.
5. Cover and refrigerate.

COURTESY AMERICAN EGG BOARD

Pickling is a time-honored way to preserve hard-cooked eggs.

for 10 minutes. Various experts I have consulted can't agree on whether or not such eggs are safe for long-term storage out of the refrigerator. A problem would arise if the pickling solution did not penetrate all the way through the eggs.

The fresher the eggs, the better. Select small and medium eggs so the pickling solution can easily penetrate. Half a dozen bantam eggs will fit into a wide-mouth pint jar. One dozen medium eggs will fit into a wide-mouth quart jar.

For the pickling solution, mix your own vinegar and spices or use the juice from prepared cucumber pickles or pickled beets. The eggs will be more tender if you pour the solution over them when it is boiling rather than letting it cool first.

Season small eggs for at least two weeks, medium eggs for at least four weeks before serving them. The acidity in the pickling solution keeps bacteria from growing, but also causes eggs eventually to deteriorate. Stored in the refrigerator, pickled eggs keep well for six months.

Oiling

Coating eggs with oil seals the shell to prevent evaporation during storage. Eggs should be oiled 24 hours after being laid so some of their carbon dioxide can escape and the whites won't take on a muddy appearance.

Into a small bowl pour white mineral oil, available at any drugstore. The oil must be free of bacteria and mold, which you can ensure by heating the oil to 180°F (82°C) for 20 minutes. Cool the oil to 70°F (21°C) before dipping the eggs.

The eggs must be at room temperature (50–70°F/10–21°C) and fully dry. With tongs or a slotted spoon, immerse the eggs in the oil one by one. To remove excess oil, place each dipped egg on a rack (such as a rack used for cake cooling or candy making) and let the oil drain for at least 30 minutes. Catch the dripping oil for reuse. Discard oil that contains debris or water, or that takes on a strange odor.

Oiled eggs may be used like fresh eggs except when it comes to cake baking — oiling eggs interferes with the foaming properties of the whites so they won't whip up as well as fresh ones. Experiments in Australia prove that oiled eggs will keep for as long as 35 days at tropical temperatures. Stored at 50°F (10°C) for eight weeks or 70°F (21°C) for five weeks, they retain their flavor better than untreated eggs.

In clean, closed cartons in a cool place, eggs dipped in oil will keep for several months. Like all eggs stored for the long term, they'll eventually develop an off flavor. The longer the eggs are stored, the greater becomes the flavor intensity compared to untreated eggs. This flavor change is pronounced in eggs stored at 34°F (1°C) for more than four months, and by six months the off flavor is unacceptable to most people.

Thermostabilization

Thermostabilization was regularly practiced by housewives during the late 19th century. Heating destroys most spoilage-causing bacteria on the shell and seals the shell by coagulating a thin layer of albumen just beneath it. When the egg cools, the coagulated albumen sticks to the egg membrane and cannot be seen in the opened egg. Unlike oiling, this method does not affect an egg's foaming properties.

Process eggs the day they are laid. Heat tap water to exactly 130°F (54°C). Use a thermometer, since the temperature is critical — the water must be just warm enough to destroy spoilage organisms but not hot enough to cook the eggs. Place eggs in a wire basket (such as a vegetable steamer or pasta cooker). Submerge the eggs in the water for 15 minutes if they are at room temperature, or 18 minutes if they have been refrigerated. Lift the basket and thoroughly drain and dry the eggs. Thermostabilized eggs will keep for two weeks at 68°F (20°C), and eight months at 34°F (1°C).

Thermostabilization and Oiling

Thermostabilization destroys bacteria and protects albumen quality. Oiling minimizes weight loss due to evaporation and preserves yolk quality. Combining the two improves an egg's keeping qualities compared to either method alone. You can thermostabilize eggs and then oil them, or combine the two procedures into one. For a combination operation, heat the oil to 140°F (60°C) and hold it at that temperature. Using a pair of tongs, rotate each egg in the hot oil for 10 minutes, then set the egg on a rack to drain. As with simple oiling, albumen foaming properties are reduced by this process, making these eggs unsuitable for cake baking.

Water Glass

Submerging eggs in water glass was the preferred method of storage during the earlier part of the 20th century. *Water glass* is a syrupy concentrated solution of sodium silicate, available from a drugstore. Its purpose is to minimize evaporation and inhibit bacteria. The water glass imparts no taste or odor and — although it causes a silica crust to develop on the outside of the shell — does not penetrate the shell.

Put eggs in water glass the same day they are laid. Candle them and eliminate any with blood spots or meat spots. As with the other processes, use only clean (not cleaned) eggs that are free of cracks. Place the eggs in a scalded glass jar with a tight-fitting lid. A 1-gallon jar will hold about 3 dozen eggs.

Combine 1 part water glass to 10 parts boiled water. If the solution is not diluted enough, it will become a gel that makes handling the eggs more difficult. Mix the solution thoroughly and let it cool. Slowly pour the cooled liquid over the eggs until the solution covers the eggs by at least 2 inches (5 cm). Do not save leftover solution. Screw the lid onto the jar to prevent evaporation. If you don't have many eggs at one time, continue adding eggs and fresh solution until the jar is full, always making sure the solution is at least 2 inches (5 cm) above the eggs.

Eggs in water glass will keep for 6 months or more if stored at 35°F (2°C).

Store the jar in a refrigerator, basement, or other cool place where the temperature is preferably not over 40°F (4.5°C). At 35°F (2°C), eggs in water glass will keep for 6 months or more. If you wish to hard cook an egg, poke a tiny hole in the big end to keep the shell from cracking as a result of the silica crust.

Even at temperatures as high as 55°F (13°C), eggs in water glass will keep for several months and be satisfactory for cooking. Under the best storage conditions, water glass causes eggs to lose their fresh flavor and take on a flat taste. The whites will eventually get thin and the yolks will flatten, making them less suitable for frying or poaching than for scrambling or using in a recipe.

ARTHUR J. MAURER, PHD

CHAPTER 8

MANAGING BREEDERS

IF YOU'RE CONTENT to keep a flock of commercial strain layers or raise an occasional batch of broilers, you needn't concern yourself with maintaining a breeder flock. But if you choose to propagate one of the seriously endangered breeds or you enjoy the challenge of trying to produce top quality show birds, you'll need a breeder flock for the production of hatching eggs.

Developing a Breed

The chicken, as we know it today, is a man-made creature. All the various breeds were developed by human design from the wild red jungle fowl of Southeast Asia. Although genetic differences distinguish one breed from another, exactly when a breed becomes a breed is purely a human invention.

By definition, a breed is commonly accepted as a family of genetically related individuals having —

- shared physical characteristics including size and shape of body, head, and comb
- shared tendencies such as broodiness or the laying of large or small eggs
- the ability to reproduce themselves.

In other forms of livestock, an animal's breed is identified by its papers. Chickens have no registry and therefore no papers, so some people would argue there are no breeds, only types. On the other hand, the papers that come with a registered animal may be based on little more than the honesty of the person who registered the animal, which is no different from accepting the verbal honesty of someone who sells you a chicken.

Some chicken keepers enjoy the challenge of developing new breeds, but most are content with preserving or improving an existing breed or variety. Some enjoy working with an established strain; others prefer to leave their mark by developing their own strains.

A strain is a family of birds having recognizable characteristics that readily distinguish them from others of their breed and variety, and the ability to transmit those characteristics to their offspring. A strain is the result of one person's (or organization's) vision, and is developed by working for many generations with a single family of birds. If you work with a group of chickens so long that other poultry people begin recognizing a bird as having come from your flock, you have developed your own strain.

Form versus Function

Every strain belongs to one of five main groups of genetic stock. Each of these groups has little contact with the others and each carries emphasis on a different set of traits. In exhibition strains, emphasis is on traits involving form. In commercial laying and meat strains, emphasis is on function. Purebred dual-purpose and sport strains combine form and function.

While their goals differ, owners of dual-purpose and sport strains share two important characteristics. First, both maintain low profiles — dual-purpose flock owners because they engage in the quiet business of home meat and egg production, sport bird owners because their business (cock fighting) is illegal in most states. Second, they share responsibility for maintaining genetic stocks having the greatest degree of sustainability.

Keepers of exhibition and industrial strains, on the other hand, have done irreparable damage to the sustainability of the genetic stocks in their care — the former by breeding for extremes in appearance, the latter by breeding for extremes in production. To understand the degree of damage done, let's look at nine characteristics that may be important, irrelevant, or detrimental within each of the five basic genetic groups. In any planned breeding program, a trait that's important to the breeder's goal is emphasized, an irrelevant trait is ignored (and thus may or may not eventually disappear), and a detrimental trait is selectively bred against.

Broodiness — a hen's desire to hatch eggs — is detrimental to egg production because once a hen starts setting she stops laying. It is irrelevant to meat production because broilers don't live long enough to reach laying age. It can be a nuisance in an exhibition strain — the more eggs you can get from a valuable hen, the more heavily you can cull the offspring. On the other hand, part of the charm (not to mention labor reduction) in keeping a purebred dual-purpose flock is having a broody hen hatch her own chicks. And among sport strains, a hen's purpose is to perpetuate her strain by laying and hatching eggs.

Fecundity, or the ability to lay eggs, is all-important in commercial laying strains but of lesser importance among meat strains. In exhibition birds, laying ability would seem important for the production of hatching eggs, but in reality the trait rates well behind characteristics of appearance. In backyard dual-purpose flocks and among sport strains, poor layers are considered freeloaders and are culled so they don't reproduce their own kind.

Fertility is important in industrial breeding flocks, but is irrelevant in production flocks. It is ostensibly important in exhibition stock, yet inbreeding small populations to focus on conformation leads to fertility loss, as does breeding for extremes of size. In a hybrid dual-purpose flock, fertility is irrelevant, but it is essential for the perpetuation of a purebred dual-purpose flock. Among sport birds, fertility is emphasized because a good brood cock makes a good battle cock and vice versa.

Foraging ability is irrelevant to industrial and exhibition stock, since both are kept in confinement — the former for reasons of labor efficiency and isolation from disease, the latter to protect plumage. In dual-purpose flocks, foraging is important as a means of cost reduction. Foraging is a significant source of exercise and nutrition for sport birds, whose owners go so far as to plant special grasses for the grazing pleasure of their stock.

Plumage color is irrelevant in commercial layers, although most strains happen to be white. In meat birds, white feathering creates a cleaner finished appearance. The greatest variety in feather color occurs among exhibition birds, yet little variation is tolerated within each color variety. Furthermore, *Standard* colors are sometimes contrary to natural tendencies, being derived by perpetually crossing different strains or even different breeds. In traditional backyard flocks and among sport strains, plumage color retains its original survival purpose — any color other than white offers camouflage for foragers and setters. Among sport enthusiasts, plumage color also identifies established bloodlines, which are called "breeds" in sport circles, but which in reality are different varieties of Old English Game.

Size is important among all five genetic stocks. For layers, small size promotes efficiency. Among meat strains, emphasis is on large size and unnaturally rapid growth. Exhibition birds must conform to the sometimes arbitrary standard sizes and weights designated for each breed, with extremes in either direction tending to mitigate against fertility and/or fecundity. Dual-purpose flocks are, by definition, mid-size as a compromise between laying efficiency and meaty flesh. Among sport strains, size relates to agility, 5 pounds (2.25 kg) being considered the ideal.

Temperament takes a backseat to other traits in the selection of industrial strains. As a result, meat birds are prone to panic and piling, and layer strains are notoriously nervous and flighty. Backyard dual-purpose flocks are bred for good temperament to enhance the keeper's enjoyment and ensure the safety of less nimble family members. Among exhibition strains, good temperament is essential, since calm birds show better than flighty ones. Game birds, too, are bred to be good-natured and gentle around people. In show and sport circles, a nervous flighty bird is referred to as *wild*, in contrast to a calm gentle bird, which is *tame*.

Type, or conformation, is unimportant to industry and to owners of backyard dual-purpose flocks. Since these breeders don't select against variations, great diversity in type exists within each given strain. At the opposite extreme, type is an essential trait in exhibition and sport strains. Since these breeders select against variations, little diversity in type exists within each strain. Exhibition promotes the greatest number of overall types, yet is the least tolerant of variations within each type.

Developing a strain with good show conformation involves inbreeding and selection, which too often leads to loss of fecundity, fertility, and/or the self-sufficient ability to forage. In addition, the quest for uniqueness in type has allowed certain traits to flourish that would otherwise inhibit survivability. Frizzledness and silkiness, for example, offer less protection from the elements than smooth, webbed feathers, and crests restrict a bird's ability to see and to get away from predators or catch mates.

Vigor is a complex trait that embodies not only resistance to disease but also adaptability to the environment, freedom from lethal genes, and the ability to produce fertile, hatchable eggs. While everyone agrees vigor is important, the trait is not high on everyone's selection list.

Meat birds need only short-term vigor, since most are dispatched by 8 weeks of age. Indeed, I've read texts offering suggestions on how to keep unhealthy meat birds alive until slaughter. Their health typically depends more on pharmaceuticals and other means of protection from disease-causing organisms than on inherent constitutional vigor.

Commercial layers were first bred primarily for their laying ability, but in later years their vigor was improved to the point that they now have genetic resistance to certain common illnesses, including Marek's disease. Like meat strains, their adaptability has been to the farm-factory environment, rather than to the great outdoors.

Exhibition birds must be hardy to stand up under the rigors of show, yet vigor is too often not a breeding priority. Purebred dual-purpose flocks tend to have inherent vigor as a result of their longevity — only the strongest individuals survive to the second year and beyond, and pass their vigor on to their offspring.

The most vigorous genetic stock is among chickens bred for sport. I once saw a particularly awe-inspiring example at a major poultry show, entered by an unwitting sport breeder. The cock — which everyone agreed was the best at the show — was the epitome of alertness and good health, was exceptionally well groomed and well tempered, yet didn't stand a chance of winning because its size and type did not conform to the *Standard*. Its superiority as a sport bird and as a representative of its species did not translate well into exhibition.

Vanishing Gene Pool

The incredible genetic diversity available within the five different groups is a phenomenon of bygone days, when flocks were raised primarily in backyards and people had more time for experimentation. The trend toward diversity was reversed with the advent of industrial farming, which concentrates genetic resources into a few strains that lay, or that grow and convert feed, uniformly well.

The old, lower-yielding breeds have been left in the hands of backyard keepers. Unfortunately, as the interests of small-flock owners shift in other directions, or we lack heirs willing to carry on with poultry, the old varieties are slipping away. In a process called "genetic erosion," the gene pool is becoming less diverse and more uniform.

Genetic erosion is not just happening among chickens but among all livestock and plant crops, as well as in wild populations that cannot withstand destruction of their natural habitat. Although genetic erosion in general is accelerating at an alarming rate, by some accounts loss of the classic poultry breeds is far worse than losses among other livestock. At the same time, no other form of livestock is so completely controlled by industry; a mere handful of companies, most of which lack long-term goals, maintain the industrial gene pool in a limited number of highly selective (and secretly guarded) strains.

But the greater the genetic diversity, the better the odds are of finding individuals with the potential to improve characteristics or resist stresses (including diseases) that change with our changing environment. A few enlightened souls, some of them within the industry, see the older breeds as a sort of insurance policy, since their traits may prove genetically useful in the future. Because poultry sperm and embryos, unlike those for cattle and other valuable stock, can't be readily preserved through freezing, the only way to perpetuate poultry genetics is through the living flocks currently kept by amateurs.

If, at some future date, commercial producers turn to backyard breeders for help, it won't be the first time. In the 1940s the broiler industry sought a broad-breasted breed to incorporate into their meat strains. They found what they needed in backyard flocks of exhibition Cornish. Ironically, at that time the Cornish had become nearly extinct — being among the many breeds that had been discarded in the scramble toward industrial egg production.

Pessimists say it's too late to save some of our endangered varieties, believing the numbers have already dwindled well below viable breeding populations. Optimists feel that any variety can still be preserved, so long as there's one pair left to breed.

Breeding Plan

In small-scale poultry circles, people who collect and hatch eggs from their flocks are divided into two camps: the so-called "propagators" or "multipliers," who emphasize quantity, and the breeders, who emphasize quality. Both hatch lots of chicks. To the propagator, the end goal *is* large numbers of chicks. To the breeder, a large number of chicks is merely the means to an end — the more chicks you have, the more heavily you can cull; the more heavily you cull, the better the genetic quality of your stock. Breeding is therefore a long-term investment. The best breeders have been at it for decades.

Breeders look down on propagators because they know that those who leave matings to chance get nowhere, genetically speaking. Indeed, if you mix chickens of several different breeds and let them mate freely, eventually their descendants will begin to look like the wild jungle fowl from which they originated.

Breeding Plan

To improve your strain and maintain its quality, follow a well thought out breeding plan that includes these steps:
- Begin with the best birds available.
- Establish a long-range goal.
- Make deliberate matings to meet that goal.
- Keep meticulous breeding records.
- Mark chicks to track their parentage.
- Cull ruthlessly.

Establishing Goals

Your breeding goal will depend on the quality of the stock you start with, compared with what you want. In selecting breeding stock, look for two things:

individual superiority and good lineage. Avoid breeding a bird with poor ancestry no matter how great it may appear.

Once you decide on a long-range goal, break it down into a series of short-term goals that will help you periodically gauge your success. Set a quality line and don't breed any bird that falls below the line. Each year, raise your quality line a little higher. Concentrate your efforts on improving one trait at a time, but not to the exclusion of others. It wouldn't do, for example, to concentrate on improving type to the exclusion of fecundity.

The most successful breeders specialize in one breed and one or only a few varieties within that breed. There's so much to know genetically about each breed that branching out too widely will dilute your efforts. After working with one breed or variety for several years, you may be ready for the challenge of taking on a new breed or an additional variety within your chosen breed.

Breeding for Show

Breeding exhibition chickens involves mating for type, since type defines breed. It also involves mating for plumage color, since in most cases color defines the variety. Select cocks and hens that closely resemble the ideal for their breed and variety as described in the *Standard*. Since every bird has both strong and weak points, avoid mating birds with the same fault. Instead, look for mates with opposite strong and weak points, but never breed a bird with great strong points if the bird also has serious faults.

In a show bird, temperament is nearly as important as type. A less typy bird that's tame will almost always win over a typier bird that's wild. Tame doesn't mean lacking in spirit, though. A good show hen is perky and likes to sing; a good exhibition cock is a show-off. Unfortunately, the more he likes to show-off, the more aggressive he is likely to be. The typiest silver Sebright I ever raised attacked me every time I fed him. Despite his fine looks, I disposed of him before he could breed more of the same.

Breeding Layers

Small populations of inbred birds invariably decline in egg production, which is why show birds generally don't lay well. On the other hand, production responds well to crossbreeding, which is why commercial layers are bred by crossing birds from different genetic lines. Breeding for commercial production is quite complex, requiring the skills of highly trained specialists. Taking a tip from them, the small flock owner can markedly improve a strain's laying ability by crossing two different inbred lines.

Good laying ability is not a highly heritable trait, meaning it is not passed directly from a hen to her daughter. A better indication of hen's worth in passing good laying ability along to her female offspring is the average rate of lay of all the hens in her family line. If the family, in general, consists of good layers, the hen is likely to pass the ability to her offspring, even if she herself is not a particularly outstanding layer. An outstanding layer, by the same token, is not likely to pass her ability to offspring if her rate of production is not typical of her family's average.

Breeding Broilers

Breeding for meat production shares much in common with breeding good layers: Both are complex, highly specialized, concentrate their genetics in a limited number of strains, and involve crossbreeding. Among commercial broiler breeders, some specialize in raising sire lines, others in dam lines, but most now work with both. Commercial broilers may carry two-part hyphenated names, such as Peterson-Cobb. The first name represents the source of the sires, the second name represents the source of the dams.

The characteristics of good meat birds — rapid growth and efficient feed conversion — are unlike laying ability in being quite heritable. Since a fast-growing bird passes the trait directly to its offspring, breeders are selected on the basis of having the greatest weight among flock mates at 8 weeks of age. While it wouldn't make sense to compete with the professionals by trying to establish a backyard broiler strain, you might use their technique to improve the growth rate of show or dual-purpose stock.

Methods

Not all the improvements that occur from one generation to the next are due to inheritance. Some may result from good management or from what is sometimes called "favorable accident." The amount of improvement attributable to inheritance is called "heritability."

In general, conformation characteristics are highly heritable, while production characteristics are not. To perpetuate traits with low heritability, breeders must be selected on the basis of family averages. To perpetuate traits with high heritability, breeders must be selected on the basis of individual superiority for those traits. Hence the evolution of two different breeding methods:

- flock breeding, which emphasizes the overall performance of a production flock
- pedigree breeding, which emphasizes individual characteristics

Flock breeding is practiced commercially and is also suitable for small-scale dual-purpose and layer flocks. Pedigree breeding is most often used for exhibition and sport strains. Pedigree breeding itself has two common methods, individual mating and pen breeding.

Individual mating can be accomplished either by mating a single cock with a single hen or by rotating one cock every two days among three or four individually penned hens. Individual matings can be tracked, or pedigreed, with certainty, since you know exactly which cock fertilized which hen's eggs.

Pen breeding involves mating a cock with a small number of hens together in one pen. It is similar to flock breeding except that a smaller numbers of hens are mated to only one cock at a time. Unless you can identify which egg comes from which hen, pen-bred birds can be pedigreed only on the cock's side. You *may* be able to identify parentage if each hen lays an egg of unique size, color, or shape. The only way to be absolutely certain of each hen's identity is to trap the hen when she goes into a nest to lay.

Linebreeding

The most common form of pedigree breeding is *linebreeding,* in which the influence of a superior sire or dam is concentrated by mating the bird to his or her best descendants. Pullets are mated to their sires or grandsires, cockerels are mated to their dams or grandams.

A good linebreeding plan includes four or more related families, starting with the best cock and four best hens available. Each family line consists of all the female offspring from one hen. The advantage to maintaining several lines is that if one of them fails to live up to your ideals, you can easily scrap it and start over with a new foundation female from within the same strain.

In general, you won't know whether the matings you have chosen will prove successful until your third year of linebreeding, when both desirable and undesirable genes become concentrated. The third year is therefore the time when novice breeders are most likely to get discouraged and quit. The more separate lines you keep, the more you decrease your chances of getting discouraged by increasing your chances that at least one line will prove successful.

Double mating is used for breeds or varieties of show stock in which the male and female differ in type or color. It involves maintaining two different breeding lines, which may or may not be related to one another. To get good show cocks, match your best male to females that tend toward cock color or type, even though they themselves aren't suitable for showing. To get good show females, mate your best females to a cock with hen color or type, even though he would not be suitable for showing. Exhibit only the top cocks from your male line and top hens from your female line.

Trapnesting

A trapnest is a nesting box fitted with a trap door. When a hen enters to lay, the door shuts and locks. The hen remains trapped in the nest until you let her out. At that time, collect her egg and mark it with the hen's identification code and the date, using a grease pencil or china marker.

To get hens used to entering trapnests, install the doors in advance and secure them in an open position. When you start trapnesting, check nests often — every 20 or 30 minutes during prime laying time. A hen confined for too long in the nest may soil or break her egg while trying to get out, and in warm weather she can suffer from being enclosed too long in a hot space.

You can either make wooden trapnest doors or buy nest fronts made of heavy-gauge wire. Although ready-made fronts are designed to fit industrial metal nests, you can easily fit one to a wooden nest by attaching it to a wooden frame and screwing the frame to the front of your nest.

FRONTS CAN BE FOLDED UP OUT OF THE WAY WHEN NOT IN USE.

FRONT IS SET AND AWAITING ENTRY BY A HEN

TRAP HAS BEEN TRIPPED BY HEN

Above, trapnest fronts are mounted on industrial (metal) nests.

ALLAN DAMEROW

The open construction of a store-bought all-wire trapnest front lets you easily see the hen in the nest and provides her with good ventilation in warm weather. The above are wire fronts fitted to homemade wooden nests.

Homemade Trapnests

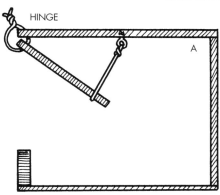 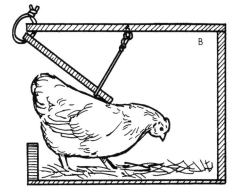

Hook Style (A and B): Lightweight wooden door is hinged with #9 wire to swing freely. A hook made of #9 wire is set to hold door open 6⅜" (16 cm) above nest floor. Hen entering the nest brushes hook loose with her back. Nest must be about 6" (15 cm) deeper than normal, so the trap door can close behind the hen.

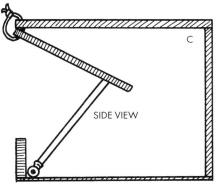 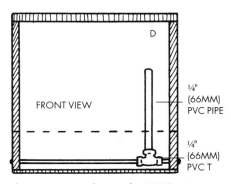

Prop Style (C and D): Lightweight wooden door is propped open by PVC pipe. A bolt or heavy-gauge wire holds the prop in. The prop falls when hen enters nest, causing trap door to close behind her.

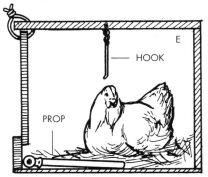

(E)With either style, the door hits the sill on the inside so the hen can't push it open. After laying her egg, she remains confined until she is set free. A few holes drilled into the door provide ventilation.

Pedigreeing

Pedigreeing is the process of keeping track of each chick's ancestry by giving every breeder an identification code and marking every hatching egg with the identity of the mating that produced it. Breeders can easily be identified with numbered wing bands or leg bands, available through poultry supply catalogs and some feed stores.

Wing bands are permanent, and one size fits all. Leg bands come in three styles: plastic spirals (which tend to break after awhile), adjustable aluminum strips, and permanently sealed aluminum strips. Permanent aluminum bands can be customized with such information as the bird's pedigree data or your name and address.

To ensure that a leg band is neither so big it falls off nor so small it binds, you'll need to get the right size for each bird. A band's size denotes its diameter in sixteenths of an inch. A #7 (⁷⁄₁₆ inch) band fits most bantams; a #9 fits light-breed hens; a #11 fits light-breed cocks, most dual-purpose cocks and hens, and heavy-breed hens; and a #12 fits heavy-breed cocks.

Pedigreeing makes no sense unless you keep meticulous mating records. Prepare a separate sheet for each mating, identifying the sire and dam (or pen number, for pen-bred birds). Then list the number of eggs you set, the number that were fertile, the number hatched, and the date they hatched.

List each chick individually so you can track it to the next generation. (Methods for identifying chicks are described on page 216). As the chicks grow, tabulate information that's important to your breeding program, such as sex, type (poor, good, excellent), color, size, temperament, age at first lay, laying ability (number of eggs per year), and the date the bird was sold or culled. If the bird was culled, indicate why.

Progeny Testing

The primary goal of pedigreeing is progeny testing, or measuring the value of breeders by the quality of their offspring. If you find a particular individual or mating that produces superior offspring, repeat the mating as often as possible. Avoid repeating any mating that produces substandard chicks.

Progeny testing is a long-term process involving keeping track of *all* the birds (not just the occasional superior bird) from each mating. You may find that certain matings produce good traits in one area (e.g., type) while others produce good traits in another area (e.g., color). If your goal is to produce superior show birds, you may wish to set up separate breeding lines for each.

Genes

In order to interpret the results of your breeding efforts and make any necessary adjustments, you have to know a little about *genes* — the hereditary units that transmit characteristics from parents to offspring. Each chick acquires two sets of genes, one from its sire and one from its dam. The genes match up into pairs of like function, for example controlling comb style or feather color.

If the paired genes are identical to each other, they are called "homozygous" (*homo* from the Greek word meaning "same," *zygous* meaning "pair"). A bird with a large number of paired identical genes is also described as homozygous. The more closely birds are related, or inbred, the more homozygous they become and the more predictable their offspring.

When the genes in a pair differ from each other, the pair is called "heterozygous" (*hetero* from the Greek word meaning "different"). The same word describes a bird with a large number of paired dissimilar genes. When heterozygous chickens are mated, or an outcross has been made, the genes in their offspring can pair off in many different combinations, making the results highly unpredictable.

Dominant versus Recessive

Each chick has a combination of dominant and recessive genes. If a dominant gene pairs up with a recessive gene, the dominant gene overshadows or modifies the recessive gene, and the dominant trait shows up in the chick.

As long as you work with hetero-zygous birds, recessive genes can remain hidden to pop up at any time. Old timers use the word "throwback" to describe a bird displaying traits that have been hidden for several generations.

A recessive trait shows up when both genes in a pair control the same recessive trait. Since a homozygous chick is more likely than a heterozygous chick to have a large number of matched pairs, the more inbred a bird is, the more likely it is to display recessive traits.

Examples of Dominant and Recessive Traits

Dominant

5 toes
feathered legs
crest
side sprigs
frizzledness

Recessive

4 toes
stubs
single comb
wry tail
silkiness

Lethal Genes

Among the recessive traits concentrated by inbreeding are lethal genes. A chick that acquires the same lethal gene from both parents dies early, often in the embryo stage. Over 50 different lethals have been identified in chickens. They are readily recognized because they are usually accompanied by such quirks as stickiness at hatch, shaking, winglessness, twisted legs, missing beaks, extra toes, and so forth.

When you mate two birds carrying the same lethal recessive, 25 percent of their offspring will display the lethal trait. The best known lethal is the so-called creeper gene carried by short-legged Japanese chickens, once prized as setting hens because their short legs keep their bodies close to the ground. Other well-known lethals are carried by Dark Cornish, New Hampshire, and white Wyandotte. Fortunately, lethal genes are relatively rare.

Two good ways to find out about lethals and other problem genes in your chosen breed are to talk to experienced breeders at poultry shows and join a breed club. Many clubs put out newsletters containing information about problems encountered by other members. Some breeds have become the subjects of entire books, which you can locate through your breed club, poultry suppliers, catalogs, listings in *Poultry Press*, or by asking experienced breeders. Even if you don't intend to show your birds, you can glean a wealth of genetic information from those who do.

Inbreeding Depression

By concentrating genes, inbreeding not only creates uniformity of size, color, and type, but also brings out weaknesses such as reduced rate of lay, low fertility, poor hatchability, and slow growth — a phenomenon called *inbreeding depression*. Inbreeding doesn't *cause* these problems but accentuates any tendency toward them.

To minimize inbreeding depression, avoid brother-sister and offspring-parent matings, and instead mate birds to their grandsires or granddams. Cull in favor of fertility, hatchability, chick viability, disease resistance, and body size. Never breed birds with any tendency toward infertility. By inbreeding gradually and culling carefully, you can actually improve such traits as laying ability and disease resistance.

Some strains are less susceptible than others to the effects of inbreeding depression. The problem isn't as great with popular breeds that have lots of varieties as it is with less popular breeds and/or those with few varieties. If you work with a breed or variety that has only a few distinct bloodlines, the time may come when you have to outcross to a different breed. If you're breeding show birds, you'll then have to work to bring them back to type.

Outcrossing

Outcrossing means increasing heterozygosity by introducing a bird that is not directly related to your bloodlines. In an inbreeding program, red flags indicating that it's time to outcross include:

- unexpected appearance of an undesirable trait
- rapid or drastic reduction in fertility, hatchability, chick viability, or general health
- continuing lack of improvement, indicating that your birds simply do not carry the right genes

When you bring in new blood, you may not see the changes you desire until the second generation. Meantime, you run the risk of introducing new weaknesses, a hazard you can minimize by crossing your strain with distantly related birds, called a *semi-outcross*. A semi-outcross is essential if you're trying to improve type, since conformation cannot be improved by crossbreeding.

To further reduce the risks of outcrossing, select new blood from a strain that isn't deficient in any of the properties you have been working to establish and that has been properly inbred. A properly inbred sire or dam is likely to be prepotent, or able to pass on its attributes to the majority of its offspring. Prepotency can only result from homozygosity.

Hybrid Vigor

Outcrossing results in *hybrid vigor*, the opposite of inbreeding depression. Hybrid vigor is a phenomenon whereby a chick is better than either of its parents. Traits with low heritability that show the greatest degree of inbreeding depression — such as size, reproductive performance, and chick viability — react the most favorably to hybrid vigor.

To realize the greatest benefits of hybrid vigor, you must maintain a high degree of heterozygosity by continually outcrossing or semi-outcrossing, which entails a constant search for new blood. If your goal is to preserve genetic diversity, not only shouldn't you cross different breeds but you shouldn't even mix established strains within a breed.

In deciding whether to create homozygosity through inbreeding or heterozygosity through crossbreeding, consider these two points:

◆ How important is predictability to your breeding program?
◆ Do you prefer to hide your birds' genetic weaknesses and hope they never surface, or force them to the surface so you can cull to eliminate them?

Sex Determination

All the genetic information that's transmitted from a chicken to its offspring is organized on chromosomes. A cock has 39 pairs of chromosomes. One pair, called the "sex" chromosomes, contain the information that determines gender. The other 38 pairs are called "autosomal" chromosomes. Like a cock, a hen has 38 pairs of autosomals, but unlike a cock, she has only one sex chromosome.

Every fertilized egg contains a sex chromosome from the cock, but a hen transmits her sex chromosome to only 50 percent of the eggs she lays. If a fertilized egg contains chromosomes from both the cock and the hen, it will hatch into a cockerel; if it contains only the single chromosome contributed by the cock, it will hatch into a pullet.

Since each egg has a 50/50 chance of containing two chromosomes, eggs hatch in approximately a 50/50 ratio of cockerels to pullets. Significant deviations from this ratio may be due either to random deaths of embryos and chicks or to sex-linked lethal genes.

You cannot determine in advance which sex will hatch from a given egg. After a chick hatches, the traditional way to learn its sex is by the Japanese method, also known as *cloacal* sexing or *vent* sexing. Accuracy depends on the skill of a trained observer in examining minor differences in the tiny cloaca just inside a chick's vent.

Sex Linkage

Since a pullet does not acquire her dam's sex chromosome, she cannot acquire any genetic information it contains. A cockerel, on the other hand, always acquires genetic information contained on its dam's sex chromosome. All characteristics that are controlled by genes on a hen's sex chromosome are called *sex linked* (all others are *autosomal*). Sex-linked characteristics include the silver color pattern (white plumage with black hackles, wings, and tail feathers), albinism, dwarfism, nakedness, barring, and late feathering.

When a hen with a certain sex-linked trait is mated to a cock without it, the trait is acquired by all the resulting cockerels, but not the pullets. Since all the pullets are like their sire and all the cockerels are like their dam, this so-called criss-cross inheritance allows the sex-linked sorting of chicks according to such things as their color or the speed of their feather growth.

Color sexing takes advantage of the sex-linked gene that controls feather color. Numerous variations are possible. If you mate a Delaware hen with a Rhode Island Red or New Hampshire cock, for example, the resulting cockerels will have the Delaware color pattern while the pullets will be solid red. If you cross a barred Rock hen with a Rhode Island Red cock, each cockerel will have a white spot on its head.

The most common commercial cross uses a silver female (such as a light Sussex) and a gold male (such as a Rhode Island Red). The female offspring are gold and the males are silver. Color sexing is most commonly used commercially for hybrid brown-egg layers. A popular example is the Hubbard Golden Comet.

Feather sexing involves crossing a slow-feathering hen (such as a Rhode Island Red) with a rapid-feathering cock (such as a white Leghorn) to get slow-feathering cockerels and rapid-feathering pullets. The chicks can be sexed with fair accuracy by the appearance of well-developed flight feathers on the wings of pullets at the time of hatch. Feather sexing is commonly used in the broiler

ALLAN DAMEROW

When you cross a barred Plymouth Rock hen with a Rhode Island Red rooster, each cockerel (right) will have a white spot on its head.

industry, where only white-feathered birds are preferred, to separate slow-growing pullets from their faster-growing brothers.

Culling

A small flock can degenerate rapidly if you make no effort to select in favor of health, vigor, hardiness, and good reproduction. In pursuit of your breeding goals, keep only your best offspring and get rid of the rest, even though they may be a large percentage of each hatch.

Just how severely you'll need to cull will depend on the quality of your foundation flock. To meet your goals, you may have to cull so strictly that you'd be better off to scrap the entire flock and start over again. In trying to reproduce a flock of New Hampshires I had in earlier years, I first acquired stock that turned out to lean too heavily toward Rhode Island Red characteristics. When I replaced them, the new hens came into lay producing white eggs and green eggs as well as the brown eggs typical of New Hamps. On the third try I finally located stock I felt was worth working with.

My experience is far from unique. Unfortunately, there isn't much you can do to ensure you're getting the stock you want except to get birds from someone who specializes in the strain you're interested in, has worked with it for a long time, and freely offers details about its background. Once you acquire your foundation stock, developing the birds to meet your goals is a matter of mating and culling.

Cull against birds that develop slowly, are not energetic, or might otherwise be described as unthrifty. Cull in favor of birds that show some improvement over their parents. Don't just visually inspect the birds, but manually examine each individual for skeletal irregularities or outright deformities. Any bird that does not measure up should go into the frying pan, the freezer, or the compost pile.

If you're raising exhibition birds, select breeders according to the *Standard*. The best age to cull cocks for show is 8 months. Pullets are at their best when they're ready to lay their first eggs, at 5½ to 7 months. Cull a laying flock for quality and quantity of eggs and the size of the hens. Choose meat birds with short shanks and compact bodies.

In all cases, cull in favor of good temperament — it's no fun raising chickens that are wild or downright mean. Although you'll hear all manner of advice on how to cure meanness, the only sure cure is to breed for good disposition.

After the first generation, start culling problem breeders so your flock will include not only good birds but also birds that transmit their good qualities to their offspring. Pay the same attention to cocks as you do to hens. Although you need fewer cocks than hens, they still make up 50 percent of your breeding flock.

Breaking Up Broodies

If you don't want your hens to set, broodiness is one of the characteristics you may wish to cull against. Broodiness, or the desire to set, is triggered by increasing day length. When a hen goes broody, she stops laying. Throughout the ages people who kept hens for eggs culled persistent broodies. As a result, hens of the lightweight laying breeds are now less apt to brood than heavier hens, and hybrids developed for commercial production are less apt to brood than purebred hens. At the other end of the scale are backyard breeds such as the cute little Cochin bantams I once had that laid only a few eggs each spring and then immediately went broody.

If you raise a rare or valuable breed, you may wish to discourage broodiness to keep your hens laying so you can hatch their eggs in an incubator or under less valuable hens. Depending on how serious the hen is about setting, you can discourage her from laying, or "break her up," by trying one of several techniques:

- ◆ Don't let eggs accumulate in the nest.
- ◆ Repeatedly remove the hen from her nest.
- ◆ Move or cover the nest so she can't get in.
- ◆ Move the hen to different housing.
- ◆ Put the hen in a broody coop.

The function of a broody coop is exactly the opposite of what its name might imply. It consists of a hanging cage, with a wire or slat floor, where the hen is housed for as long as it takes to break her up, usually one to three days. The longer the hen has been allowed to be broody, the longer she'll take to come back into production. A hen that's broken up after the first day of brooding should begin laying in 7 days; a hen that isn't broken up until the fourth day won't start laying for about 18 days.

Hens, like people, don't always react as you expect them to, and a persistent broody may continue no matter what you do. In that case, if you are just as determined that a hen not brood as she is to brood, your only remaining option is to cull. Sometimes a hen will be so insistent on brooding that she'll stay on the nest even if she has no eggs to hatch, eventually dying of starvation and thus culling herself.

Breeder Flock Health

Hens in poor health may fail to supply their eggs with all the nutrients needed for the development of healthy chicks, may pass along toxic substances, or may transmit diseases to their offspring. Diseases can be transmitted through hatching eggs in three ways:

◆ The yolk may become infected when an egg is formed in the body of an infected hen.

◆ In rare cases, an infected cock may fertilize an egg with infected sperm.

◆ Most commonly, bacteria penetrate a shell that becomes contaminated while the egg is being laid or after it lands in a dirty nest.

An infected embryo either dies in the shell toward the end of incubation or hatches into an infected chick. Infected chicks may spread disease to healthy chicks.

To ensure a good hatch, and to give chicks the best start in life, take special care to maintain a healthy breeder flock. A hen in the peak of health passes along to her offspring maternal antibodies that protect her chicks from diseases in the flock's environment for the first several weeks after the hatch. Strong chicks then develop their own antibodies for continued immunity.

If you find it necessary to bring in new blood for your breeding flock, acquire chicks and raise them yourself. You'll run less risk of introducing a disease into your established flock.

Fertility

When a cock mates with a hen, sperm travel quickly up the oviduct to fertilize a developing yolk. If the hen laid an egg shortly before, the mating will likely fertilize her next egg. How many additional eggs will be fertilized by the mating varies with the hen's productivity and breed. In general, highly productive hens remain fertile longer than hens that lay at a slower rate, and single-comb breeds remain fertile longer than rose-comb breeds — as long as a month, but that's pushing your luck. The average duration of fertility is about ten days.

If you switch cocks, future eggs are more likely to be fertilized by the new cock than by the old one. To be reasonably certain eggs are fertilized by the new mating, wait at least two weeks before collecting them for hatching.

Although an egg must be fertilized in order to hatch, not all fertilized eggs do hatch. Eggs fail to hatch for a variety of reasons, not all of them easy to determine. One possible reason is that the embryo died before incubation began. This phenomenon, commonly known as "weak fertility," may have nothing to do with fertility at all, but may be due to other deficiencies within the egg.

Like all reproductive qualities, fertility has low heritability. Aside from problems related to inbreeding depression, management factors (rather than inheritance) are more likely to play a significant role. The many possible reasons for low fertility include the following:

- The flock is too closely confined.
- The weather is too warm.
- Breeders (both cocks and hens) get fewer than 14 daylight hours.
- The cock has an injured foot or leg.
- Breeders are infested with internal or external parasites.
- Breeders are diseased — especially troublesome for hatching eggs are chronic respiratory disease, infectious coryza, infectious bronchitis, Marek's disease, and endemic (mild) Newcastle disease.
- Breeders are too young or too old.
- Breeders are stressed due to excessive showing.
- Breeders are undernourished.
- The cock-to-hen ratio is too high or too low.

Battle of the Sexes

The ideal cock-to-hen ratio is influenced by the cock's condition, health, age, and breed. On average, the optimum ratio for heavier breeds is 1 cock per 8 hens, although a cock in peak form can handle up to 12 hens. The optimum ratio for lightweight laying breeds is 1 cock for up to 12 hens, yet an agile cock may accommodate 15 to 20. The mating ratio for bantams is 1 cock for 18 hens, although an active cockerel might handle as many as 25. An older cock or an immature cockerel can manage only half the hens of a virile yearling.

If you have too many cocks, fertility will be low because the cocks will spend too much time fighting among themselves. If you have too few cocks, fertility will be low because the cocks can't get around to all the hens. If you have only one cock and more than half a dozen hens, the cock will likely favor some hens and ignore the others.

Since cocks tend to play favorites, it pays to switch them periodically. The more cocks you have to switch, the less likely you are to experience inbreeding problems; it's far better to have chicks sired by several different cocks than hatch the same number of chicks all with one sire. And by keeping extra cocks, you won't be left high and dry if your favorite rooster becomes incapacitated during the breeding season.

If you need only one cock at a time, house the extras in separate pens or cages. A friend of mine learned this the hard way. He knew one rooster was enough for his dozen hens, but he kept a second one as insurance. The two cocks fought so incessantly that feeders and perches soon became painted with blood, and the hens produced no fertile eggs. The exasperated fellow sold one cock to a neighbor, whose coop was within earshot. Whenever the neighbor's cock crowed, the home boy flew up to the rooftop to issue a defiant answer. One day while jumping down from the roof, the cock broke his foot, abruptly ending my friend's breeding season. This unfortunate experience would have been avoided if the fellow had alternately penned each cock alone and given it a turn with the hens.

Ironically, cocks are less apt to fight if you keep three or more, instead of just two. If you use several cocks for flock breeding, keep them in two or more groups and rotate the groups, rather than individuals — the less you disturb their peck order, the less fighting will occur. If you lose one cock out of a group during the season, put up with a possible slight drop in fertility rather than replacing him and running the risk that peck-order fighting will cause fertility to plummet.

Female/Male Mating Ratios

	Optimum	Maximum
Bantam	18	25
Light breeds	12	20
Heavy breeds	8	12

Treading

A cock may pull out a hen's feathers while treading during mating. A hen with missing feathers on her back has little protection from the cock's claws during future matings, and as a result may be seriously wounded. If a hen has her sides sliced up by a rooster's toenails, isolate the hen and treat the wounds until they heal.

To protect hens from particularly brutal roosters, house the cocks in private coops and let each run with the hens for only a day or two per week. Another solution is to clip the cocks' toenails. As an extreme measure, well before the breeding season cut off the tips of the cock's toes with pruning sheers, cauterize the wounds with a hot soldering iron, and isolate the cock for a week until the toes heal. While they heal, keep the toes clean by washing them with hydrogen peroxide, applying Neosporin, and wrapping the toes with a gauze bandage.

As a less extreme measure, dress each hen in an apron or saddle made from two pieces of canvas or cotton denim stitched together and fitted with elastic straps. Apply the saddle when the hen's feathers start disappearing, not after she's already wounded.

Make a preliminary pattern from the legs of old denim jeans and keep adjusting it until it fits properly. If the saddle is too tight it will chafe, rub off breast feathers, injure the hen's wings, or strangle the hen. If it's too loose it will flop to one side and be useless.

To dress a hen, put her head through the center opening between the two elastic straps, then put one wing through each of the other openings so a strap runs beneath each wing. When first dressed, the hen will try to back away from the saddle (please refrain from wounding her dignity by laughing), but soon enough she'll get used to it. To readily identify each saddled hen, put her leg band number on the saddle using paint, embroidery, machine stitching, or iron-on patches.

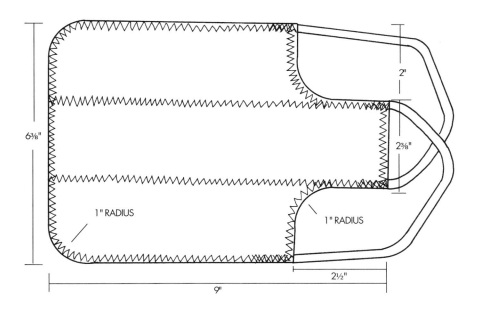

To make a saddle proportioned for a Rhode Island Red hen, you will need two pieces of denim or canvas 6⅜" x 9", two pieces of ¼" elastic, and 8½" of thread.

Breeder Flock Age

In general, expect maximum fertility and hatchability from mature cockerels and pullets. Most cockerels reach sexual maturity around 6 months of age, although early maturing breeds may be ready to mate sooner while late maturing breeds may not be ready until 7 to 8 months. Comb development is the best indication of maturity.

You can start collecting hatching eggs when pullets are about 7 months old and have been laying for at least 6 weeks. Eggs laid earlier tend to be low in fertility. The few that hatch are likely to produce a high percentage of deformed embryos (perhaps due to the relatively small yolks of early eggs). As time goes by, fertility and hatchability improve, leveling out by the sixth week.

After about 6 months of lay, fertility and hatchability begin to decline — most gradually among bantams, and more rapidly among the heavier breeds than among the lighter breeds. In industry, broiler breeders are kept for only 10 months or less, compared to 12 months or more for layer breeders.

A much-debated question among small-flock owners is whether it's better to hatch eggs from hens that are over or under 2 years of age. If your goal is to improve the health and vigor of future generations, hatch from older hens. Two-year-old hens that are laying well must be relatively disease resistant and are likely to pass that resistance to their young. Furthermore, older birds tend to be the more valuable breeders, since they have proven their ability to pass desirable traits along to their offspring (less desirable breeders having long since been culled).

On the other hand, there's a slight but significant decline in hatchability after a hen's first year, and the decline continues as the hen gets older. After the second year, you'll see a greater percentage of early embryo deaths and failure of full-term embryos to emerge from the shell. If you breed old birds, take special care to keep them stress-free, for example, by leaving the show circuit to the younger generation.

Showing and Fertility

The frequent showing of breeders can result in poor fertility and/or inferior chicks. Birds become stressed by travel, inconsiderate spectators, peculiar feed and feeding schedules, and perhaps lack of water due to oversight or simply because the birds don't like the taste. Lack of water is particularly a problem with hens in lay.

Hens, in general, are more greatly stressed by showing than cocks, and older birds of either sex are more strongly affected than younger ones. Keep valuable breeders away from the showroom and the consequent exposure to stress and potential disease, not to mention the possibility of theft. Alternatively, minimize your risks by hatching a hefty batch of chicks before showing your breeder stock.

Breed-Related Fertility Problems

Low fertility can be breed related as a result of either hereditary or mechanical problems. The most common hereditary trait that influences fertility is comb style.

For some reason, single-comb breeds tend to have higher fertility than rose-comb breeds, and the sperm of single comb or heterozygous rose-comb cocks live longer than the sperm of homozygous rose-comb cocks. So, although purists insist that the occasional single-comb chick commonly hatched from rose-comb parents should not be used for breeding, old-timers keep them in low numbers to ensure heterozygosity and good fertility.

Mechanical problems that affect fertility include:

- ◆ Comb size — Breeds with large single combs have trouble negotiating feeders with narrow openings, and the resulting nutritional deficiency affects fertility. Large combs may also suffer frostbite, resulting in reduced fertility or even sterility.
- ◆ Heavy feathering — Brahmas, Cochins, Wyandottes, and other heavily feathered breeds have trouble mating unless their vent feathers are clipped back.
- ◆ Foot feathering — Booted bantam cocks and males of other breeds with heavy foot feathering have trouble getting a foothold when treading hens.
- ◆ Crests — Houdans, Polish, and other heavily crested cocks may not see well enough to catch hens unless their crest feathers are clipped.
- ◆ Rumplessness — Araucanas (and occasionally birds of other breeds) have no tail; as a result, the feathers around their vents can't separate properly for mating.
- ◆ Heavy muscling — Cornish cocks and other heavy-breasted males have trouble mounting hens due to the wide distance between their legs.

Artificial Insemination

Artificial insemination is used primarily by breeders of Cornish birds, by exhibitors who wish to keep hens in show condition, and by owners of valuable cocks that tend to be shy or otherwise low in sex drive. The technique is not difficult, but like everything else takes practice — on the part of both you and the birds involved. As you might expect, tame birds are easier to work with than wild ones.

House cocks you wish to collect, or "milk," in separate coops, away from other birds but preferably within sight of hens. Hens may be kept together, but should be separated from other cocks if you wish to pedigree the chicks.

Collection

For semen collection, or "milking," you'll need a small glass or cup about the size of an eye cup and a one cc eye dropper or syringe (without a needle). The procedure is easiest if you have a helper to collect the semen while you do the milking. Hold the cock in the palm of one hand, with your fingers toward the back end and his legs secured between them. With the fingers, massage the bird's abdomen gently, but rapidly and continuously. With the other hand, vigorously stroke his back above the testes toward the tail.

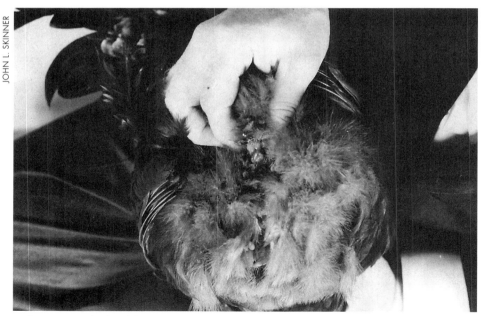

JOHN L. SKINNER

A bantam cock should give you at least 0.2 cc of semen, a large breed 0.4 cc.

After several strokes, move your free hand so the palm pushes the tail feathers up out of the way while your thumb and index finger apply pressure on both sides of the vent, ready to squeeze (but not yet). Stop massaging the abdomen with your other fingers and press upward.

As soon as the cock's organ appears, gently squeeze out the creamy-white semen while your helper collects it in the cup. You should get at least 0.2 cc of semen from a bantam and 0.4 cc from a large bird, although you may get as little as 0.1 cc or as much as 0.8 cc from either. If the cock is fertile, the concentration will range between 3 and 4 billion sperm per cc.

Should the semen become contaminated with fecal matter or chalky white urates, discard it and try again another day. If the cock persists in discharging feces at the same time as semen, withhold feed and water for 4 to 6 hours before collection. A healthy cock can be collected once every 3 days.

Insemination

Be ready to inseminate hens right away, since fresh semen loses its fertilizing capacity after about an hour. Have the semen ready in the syringe or eye dropper. If you are inseminating only one hen, use all the semen from one cock. If you inseminate more than one hen, use 0.1 cc for each.

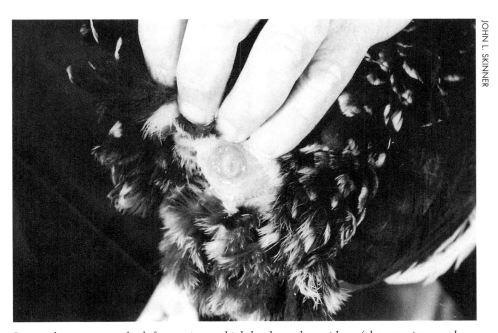

JOHN L. SKINNER

Insert the semen in the left opening, which leads to the oviduct (the opening on the right goes to the intestine).

Inseminate only a hen that has been laying, otherwise you might injure her. Feel her abdomen to make sure a finished egg isn't coming through to block the passageway. Hold the hen and massage her the same way you did the cock. When you apply pressure on her abdomen, her vent will open up. You'll see a fold of skin at the opening to the oviduct on the hen's left (the opening on the right comes from the intestine).

Your helper should quickly insert the syringe 1 inch (2.5 cm) into the oviduct and inject the semen. As soon as the semen is injected, release your pressure on the hen's abdomen to let the vent close so the semen is drawn in.

You should start getting fertilized eggs two to three days after the first insemination. Early in the season, inseminate hens at least every 7 days, preferably every 5 days to ensure good fertility. As the season progresses and fertility drops, inseminate more often.

Breeder-Flock Housing

Breeder-flock housing plays an important role in the fertility and hatchability of eggs. Facilities with lots of environmental variety help give a cock privacy to mate undisturbed by other cocks. Excessive fighting among cocks may be a sign of poor facility design, since cocks that are lower in peck order have no place to hide from dominant birds.

Floor space for the breeder flock should offer a minimum of 3 square feet (300 sq cm) per bird for large breeds, 2 square feet (200 sq cm) for smaller breeds, and 1.5 square feet (150 sq cm) per bantam. Include at least one nest for every four hens and frequently change nesting material.

Housing should protect the flock from extremes of climate, since sudden changes can cause a decrease in laying or fertility. During the early part of the season (late winter), provide lighting not only for warmth but to stimulate egg and semen production. In addition to controlling lights for egg production (as described on page 124), you can improve fertility by exposing cocks to 14 hours of light, artificial and daylight combined, for 4 to 6 weeks prior to collecting eggs for hatching.

Outside disturbances can upset a flock and interfere with mating. Make sure your birds are well protected from predators, unruly dogs, and children who run near pens or tease the birds. You may also have to protect your flock from ignorant adults. I once reprimanded a grown man who was having fun watching his dog run back and forth in front of my breeder pen. In total amazement, the man insisted the dog was having a great time without harming my birds. Chickens, this poor misguided fellow insisted, *always* flap and squawk.

Feeding Breeders

Assuming your breeder flock is healthy and free of both internal and external parasites, good nutrition is the greatest factor in promoting fertility and hatchability. Poor breeder-flock nutrition can arise because rations are:

- poorly balanced
- insufficient
- nutritionally deficient

The main cause of poorly balanced rations is feeding breeders too much scratch grain or other treats. To ensure the proper protein/carbohydrate balance, reduce your flock's grain ration about a month before the hatching season begins.

See that your breeders get enough to eat for their size, their level of activity, and the time of year. Feed either free choice or often so those lowest on the totem pole will get a turn at the hopper. Examine each bag of feed to make sure it isn't dusty, moldy, or otherwise unpalatable, causing your chickens to eat less.

Periodically weigh a sampling of both cocks and hens — a weight loss of over 10 percent can affect reproduction. Underfed cocks produce less semen; underfed hens don't lay well.

A hen's diet affects the number and vitality of her chicks and the quality of carry-over nutrients the chicks continue to absorb for several weeks after they hatch. Nutritional deficiencies that may not produce symptoms in a hen can still be passed on to her chicks. Feeding the breeder hen therefore includes feeding her not-yet-hatched chicks.

The same ration that promotes good egg production won't necessarily provide embryos and newly hatched chicks with all the elements they need to thrive. Lay ration contains little animal protein and too little vitamins and minerals for proper hatching-egg composition and high hatchability. Feeding lay ration to a breeder hen may result in a poor hatch or nutritional deficiencies in her offspring. The older the hen, the worse the problem becomes.

To improve your hatching success, feed your flock a bona fide breeder ration, starting two to four weeks before you intend to begin hatching. If you're lucky, you'll find breeder ration at your feed store. Be sure it's fresh, and use it within two weeks after it was mixed — even if you feed your flock the best breeder ration in the world, nutritional deficiencies will result if the feed is stored so long that fat soluble vitamins are destroyed by oxidation.

In areas where breeder ration is not available, the closest alternative may be game-bird ration. If you can't find either, six weeks before you begin collecting hatching eggs supplement rations with a handful of dry cat food two or three times a week, and add a vitamin/mineral supplement to the drinking water.

Supplements

Excessive embryo mortality during mid and late hatch can be a sign of low vitamin levels in the breeder ration. If you have reason to question the freshness of the ration you use, feed your flock natural supplements or add vitamins to their drinking water prior to and during the breeding season.

Vitamin A is essential for good hatchability and chick viability. It comes from green feeds, yellow corn, and cod liver oil. If you use cod liver oil, keep it fresh by mixing it into rations at each feeding.

Vitamin D is related to the assimilation of calcium and phosphorus needed for egg production. Deficiency causes shells to become thin. Since an embryo takes calcium from the shell, thin-shelled eggs may produce stunted chicks. Two signs of deficiency are a peak in mortality during the nineteenth day of hatch and chicks with rickets. Vitamin D comes from cod liver oil and sunlight.

Vitamin E affects both fertility and hatchability. It comes from wheat germ oil, whole grains, and many fresh greens.

Riboflavin, one of the B vitamins, is often deficient in poultry rations, resulting in embryo deaths in early or mid-incubation, depending on the degree of deficiency. Chicks that do hatch may have curled toes and may grow slowly. Riboflavin comes from leafy greens, milk products, liver, and yeast.

Calcium deficiency can cause thin-shelled eggs and reduced laying. Calcium excess can reduce hatchability. Supplying a calcium supplement free choice rather than mixing it into rations allows for differences in the needs of individual hens. For fast results, use agricultural limestone (*not* dolomitic limestone, which is detrimental to egg production and hatchability). Crushed oyster shell is a good long-term supplement.

The touchy problem of providing breeders with adequate nutrition will be compounded if you hatch early in the season, before your hens have access to fresh greens. If you live in a northern area, you can measurably improve your hatching success by delaying incubation until your birds have access to forage and plenty of sunlight.

CHAPTER 9

INCUBATION AND HATCHING

ASSUMING YOUR BREEDER FLOCK is healthy, is properly nourished, and contains the right proportion of cocks to hens, successful hatching depends on:

- culling eggs as rigorously as you cull breeders
- storing eggs for a minimum amount of time at optimum temperature and humidity
- maintaining excellent sanitation during egg storage and incubation
- turning eggs during incubation (and during prolonged storage)
- properly controlling temperature, humidity, and air flow during incubation

Egg Collection

Wouldn't it be nice to save incubator space by hatching eggs that mostly produce pullets? Unfortunately, despite old wives' tales to the contrary (e.g., that cockerels hatch from elongated eggs or those over which keys will swing longitudinally), no way has yet been found to determine the sex of a fertilized egg or a developing embryo without destroying the egg or embryo. This may soon change, with the poultry industry looking for a way to use biotechnology to sex eggs prior to incubation.

Until that day comes, the best you can do is ensure high-quality chicks by selecting eggs that are the proper size, shape, and color for your breed. Small eggs laid by pullets or old hens will give you smaller, less vigorous chicks. Excessively large eggs may hatch into bigger chicks, but their growth rate will likely be inconsistent. Eliminate eggs that are round, oblong, or otherwise oddly shaped, and those with shells that are wrinkled, glassy, or abnormal in any other way.

Candle the eggs to eliminate any with two yolks, which rarely hatch. If you're raising replacement layers rather than show birds, also candle for blood spots, which can be hereditary. During candling you might find hairline cracks, showing up as white veins in the shell. Don't incubate cracked eggs, since cracks open the way for bacteria to enter.

Incubation temperature and humidity provide an ideal environment for the growth of bacteria, making nest and egg-storage hygiene essential. Bacteria can cause eggs to explode during incubation or can infect developing embryos, resulting in mushy chick disease (omphalitis). When possible, avoid incubating dirty eggs. If the eggs are extremely valuable, wash them in water that's warmer than the eggs, adding a little chlorine bleach as a sanitizer.

If eggs consistently get dirty in the nest, reconsider your nest design. One option is to build nests that allow eggs to roll to the outside for collection. Another option is to use Astroturf pads, which are supposedly more hygienic than organic litter. I've found, however, that pads are difficult to clean. If you choose to use them, keep extra ones on hand so you'll have enough to line all

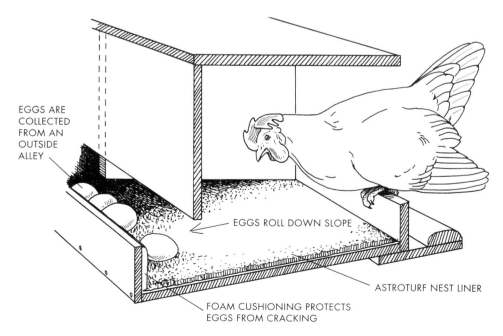

EGGS ARE COLLECTED FROM AN OUTSIDE ALLEY

EGGS ROLL DOWN SLOPE

ASTROTURF NEST LINER

FOAM CUSHIONING PROTECTS EGGS FROM CRACKING

This nest is designed to ensure clean eggs.

the nests while dirty pads are being soaked, scrubbed, disinfected, and dried in the sun. Introduce pads at the pullet stage, since hens that have grown used to nesting in natural bedding may shun Astroturf.

Collect eggs at least three times a day to minimize their contact with dirty surfaces and to keep them from getting chilled or overheated, thus reducing their hatchability.

Egg Storage

An egg need not be rushed into the incubator the moment it is laid. Egg storage is, in fact, a natural part of the incubation process. In nature, just-laid eggs go dormant to give a hen time to accumulate a full setting before she starts to brood. When she selects a place to make her nest, the hen instinctively seeks out conditions that are best for storage and incubation. When you gather and store hatching eggs, your job is to try to duplicate those conditions.

Store the eggs out of sunlight in a cool, relatively dry place, but not in the refrigerator. The best storage temperature is 55°F (12°C).

Humidity should be low enough to prevent moisture from condensing on the shells, which would cause any bacteria present to multiply more rapidly. Excessive humidity also attracts damaging molds.

Excessive dryness, on the other hand, increases the rate at which moisture evaporates through the shells. The less moisture that evaporates from eggs during storage, the greater their future hatching rate. The speed of evaporation varies from one breed to another. Small eggs, such as those laid by bantams and jungle fowl, have a relatively large surface-to-volume ratio and therefore evaporate more quickly than larger eggs. Late summer eggs of any size have thinner shells that allow more rapid evaporation than occurs in early season eggs. To minimize evaporation of eggs that must be stored longer than six days, seal cartons in plastic bags or, better yet, wrap eggs individually in plastic wrap.

Store eggs in clean cartons. Hard plastic egg cartons (of the sort backpackers use) can be easily disinfected and reused from year to year; recycled styrofoam or cardboard cartons from the grocery store will accumulate bacteria over time.

Place eggs in the carton with their large ends upward to keep the yolks centered within the whites. If the eggs will be stored for longer than six days, keep yolks from sticking to the inside of the shell by tilting the eggs from one side to the other. Instead of handling eggs individually, slightly elevate one end of the carton one day, and the opposite end the next day.

The longer eggs are stored, the longer they will take to hatch. Also, their ability to hatch decreases with time. You can store eggs for up to six days

Broodiness By Breed

Will Brood	Rarely Brood	Sometimes Brood
Australorp	Ancona	Aseel
Brahma	Andalusian	Dorking
Buckeye	Campine	New Hampshire
Chantecler	Hamburg	Plymouth Rock
Cochin	Houdan	Rhode Island Red
Java	Leghorn	Shamo
Langshan	Minorca	**Unsuccessful as Broodies**
Old English Game	Polish	
Orpington	Sicilian Buttercup	
Silkie	Spanish	Cornish
Sumatra		Hybrid layers
Sussex		Jersey Giant
Wyandotte		Malay

without noticing a significant difference. For each day thereafter, hatchability will suffer by approximately 1 percent.

Natural Incubation

Natural incubation has one big disadvantage over artificial incubation — the hen, not you, decides when it's time to hatch eggs. On the plus side, natural incubation is less time-consuming for you: You don't have to worry about the electricity going out, a thermostat malfunctioning, or someone tripping over the cord and pulling the plug; the hen provides automatic temperature and humidity control, and she turns the eggs at exactly the right intervals.

When a hen accumulates a clutch of eggs and goes broody, her pituitary gland releases *prolactin*, a hormone that causes her to stop laying. Because laying stops when setting starts, the brooding instinct has over the years been selectively bred out of layer strains. The best laying breeds are therefore the least likely to set.

As a general rule — to be broken at a hen's merest whim — Continental and Mediterranean breeds are nonbrooding, American and English breeds tend toward broodiness (some strains more strongly than others), and Asiatics make exceptional broodies. Some hens, including Cornish, Malay, and Old English Game, are apt to brood but don't always make good mothers.

Among bantams, Cochins, Silkies, and Wyandottes make outstanding broodies. Feather-legged breeds in general are likely to set, but they aren't always successful because their leg feathers flick eggs out of the nest or bowl over newly hatched chicks. To solve the problem, clip the broody's feathers close to her leg.

In the same way that broodiness and strong mothering instincts were bred out of certain strains, these traits can be improved in your flock by culling in favor of them. As a sex-linked characteristic, brooding is transmitted to pullets through their sire rather than through their dam.

Broody Signs

Just because a hen is sitting on a nest doesn't necessarily mean she's setting. She may still be thinking about the egg she just laid or she may be hiding from some bully that's higher in the peck order. To test a hen for broodiness, gently reach beneath her and remove any eggs you find there. If she runs off in a hysterical snit, she's not broody. If she pecks your hand, puffs out her feathers, or growls, things are looking good. Within two or three days she'll likely settle down to serious business.

It's not true that you can encourage a reluctant hen to set by enclosing her with a pile of eggs. She's more likely to scatter and break the eggs while trying to get out. You *can,* however, smoke out a potential broody by letting eggs accumulate in the nest. In case this takes long enough for real eggs to rot, use plastic or wooden eggs from a hobby shop. Leave half a dozen fake eggs in the nest until one of your hens shows interest. After you're sure she's serious, replace the faux eggs with hatching eggs.

Clucking is one sure sign of broodiness. Many broodies won't cluck until their eggs are ready to hatch, but some start clucking almost as soon as they start setting. The hen's habit of clucking to reassure her chicks has led to the nickname "clucker" for a broody hen. Some broodies go so far as to practice for motherhood by ruffling their feathers and clucking whenever they're off the nest.

Brooding Nests

It's a good idea to separate a setting hen from the rest of the flock, since another hen may enter her nest while she's off eating, causing her to resettle in the wrong nest. If she does manage to hatch out a brood, other chickens may kill the fuzzy intruders, much as they would kill a mouse or a frog that wanders into their yard.

Move the hen after dark, or she'll likely try to get back to her old nest. Give her a few less valuable or fake eggs until you're sure she's serious, then replace them with hatching eggs. They needn't be the hen's own eggs — she won't know the difference.

If you have more than one hen brooding at a time, separate them from each other. Otherwise the hens may accidently switch nests, shortening the incubation period for one hen and prolonging it for the other. Both setters may follow the first chicks that hatch, leaving the remaining eggs to chill. Even if each hen tends to her own chicks, one may capture the feed and watering stations for her brood and keep the other's chicks away. Avoid all these problems by keeping broodies in separate brooding compartments.

A suitable brooding nest is darkened, well ventilated, and protected from wind, rain, and temperature extremes. It should be 14 inches (35 cm) square and at least 16 inches (40 cm) high, with a 4 to 6 inch (10 to 15 cm) high lip at the front to hold in nesting material. Clean, dry wood shavings, especially cedar, make the best nesting material. Straw and hay are less suitable because they readily mold, possibly infecting developing embryos or newly hatched chicks.

Be sure both litter and hen are free of lice and mites, which can make a hen restless enough to leave the nest. These parasites can take enough blood to kill a setting hen and/or her newly hatched chicks. The use of cedar shavings discourages parasites, as does sprinkling a poultry-approved insecticide in the nest before adding litter. In the days before chemicals became widely available, poultry keepers put tobacco in the nest as an insecticide for broodies.

As the hen sets, she develops defeathered brood patches on the sides of her breast that serve two purposes: to bring her body warmth closer to her eggs, and to keep the eggs from drying out by lending them moisture from her body. Give her a hand by hollowing the nesting material at the center, making the bowl small enough to keep eggs from rolling away beneath her but big enough so they won't pile up and crush each other. Broken eggs attract bacteria and ants, so check at least once a week and, if necessary, remove egg-soaked nesting material and wash dirtied eggs in warm water.

Any time you handle a broody hen, first gently raise her wings. She may be holding an egg between a wing and her body. The egg might break — and might break other eggs, as well — if it drops into the nest as you lift the hen.

Eggs are also likely to break or roll away from the hen if she's trying to cover more than she can handle. Most hens can cover 12 to 18 eggs of the size they lay. A banty can hatch only 8 to 10 eggs laid by a large hen, while a larger broody might cover as many as 24 banty eggs. Watch the nest for the first few days and remove any egg that peeps out around the edges so it doesn't get rotated back in, leaving a different egg out in the cold and spoiling both.

Broody Management

A setting hen will get off the nest for a few minutes each day to eat. Most hens hold their droppings until they leave the nest. In case of an accident, keep droppings solid by feeding broodies scratch grain instead of lay ration.

A setting hen should start out in the peak of health with a good layer of body fat. She won't eat much during the 21 days she's on the nest and will need fat reserves to see her through. Some broodies eat so little they waste away to practically nothing by the time their chicks hatch. A persistent hen that for some reason fails to hatch a brood may eventually starve to death.

Put food and water near the nest, but don't worry if the hen doesn't eat for the first few days. After that, she should get off the nest for 15 or 20 minutes at about the same time each day to eliminate, grab a few kernels of grain, maybe take a dust bath, then zip back onto the nest. If your hen doesn't seem to be getting off the nest to eat, encourage her by either lifting her off the nest and putting her down near the feeding station or sprinkling a little scratch in front of the nest to pique her interest. If she remains indifferent, chances are she's been eating when you're not around.

After the sixteenth day, do not disturb the broody or her eggs. See that she has plenty to eat and drink, then contain your impatience until the little peepers pop into the world. While the chicks are hatching, keep an eye on things but don't interfere unless your help is absolutely necessary.

Normally all the chicks will hatch within a few hours of each other. The hen will remain on the nest for another day, maybe two, before venturing forth with her brood.

If for some reason the hatch is slow, the hen may hop off the nest to follow the earliest chicks, leaving the rest to chill and die. In that case, gather up and care for the first chicks, returning them to the hen after the hatch is complete.

Occasionally a hen will be so horrified by the appearance of interlopers beneath her that she'll attack the little fuzzballs. Be ready to rescue the chicks and brood them yourself.

If you're hatching particularly valuable chicks, you might wish to brood them yourself in any case. Some breeders of exhibition birds jump the gun by moving term eggs to an incubator for the hatch. That's tricky business, though, since any delay while moving the eggs, or an incorrect setting of the incubator, can ruin the hatch.

Foster Broodies

Some breeders feel it's surer to hatch naturally, but safer to brood artificially. Others find it just the opposite — the difference being in their facilities' setup and management style. If you're away from home much of the time, you'll likely find it less trouble to let a hen do the brooding, even when she didn't hatch the chicks herself. A setting hen can often be tricked into raising any chicks she finds under her — even those of another species. Mixing different species usually doesn't work but is worth a try in an emergency.

A hen is most likely to accept foster chicks if she's been seriously brooding for a couple of weeks but not necessarily the entire 21 days. By the same token, the chicks must be no more than a day old and still receptive to accepting (or "imprinting") a new mom. To increase the chances that both parties accept each other, slip the chicks under the hen at night. In the event things don't work out as you had planned, be prepared to gather the chicks back up and raise them yourself.

German researcher Erich Baeumer once placed a microphone near a nest of hatching eggs and listened on his radio amplifier. He was surprised to hear chicks peeping from within their shells, while the hen comforted them with a variety of sounds. Since a hen and her brood make initial contact through sound rather than sight, a foster broody is more likely to accept chicks slipped under her at night. By morning both parties will have developed an amiable relationship through sound.

Since a hen can successfully mother up to three times the number of chicks she hatches, you might wish to artificially incubate additional eggs with the intent of having your hen brood the resulting chicks along with her own. This ploy works only if all the chicks are the same age. Again, slip the freshly hatched incubated chicks under the hen at night. If you're moving the hen and her brood from the nesting site to a brooding pen, place all the chicks in the pen first so they'll get mixed together before you introduce the hen.

Brooding Pens

A hen can't always find a safe place in which to brood her chicks. In my early years of keeping chickens, I once let a broody make her nest in the layer house. When the chicks were ready to leave the nest, they fell into the droppings pit and had to be rescued. Another hen stole her nest in the hayloft. One morning I found chicks wandering around the barn peeping miserably while the oblivious hen sat happily on her nest in the mow. To prevent such mishaps, either move the hen to a brooding pen from the start, or wait until after the hatch to confine her to a brooding pen where she can safely raise her chicks.

A brooding pen is a small coop where the hen and her chicks will be safe from the elements and from predators. No single specific design is better than any other. You might adapt a small dog house, a cleaned oil drum, or an old camper shell. If you make the coop from scratch, it needn't be more than 3-feet

(90 cm) square. Be sure the floor, walls, and roof are rat proof, and install screened vents near the top for good air flow. Add a door that opens onto a small enclosed run. If the coop is outdoors, include a slatted awning for shade.

Never let a hen and her brood just wander around the yard, or the chicks may fall prey to house cats, hawks, and other predators. They might get chilled in damp grass or get lost and not be able to find their way back. Since hens can't count, once a chick is out of earshot a hen has no way of knowing the little one is missing.

Feed the chicks starter ration and put the hen back on lay ration. Place the hopper of lay ration high enough off the ground so the chicks can't reach it — the large amount of calcium required by hens will damage a chick's kidneys. Place the starter ration in a chick feeder with holes too small for the hen to peck into and anchor the feeder down in some way so she can't overturn it

Brooding coop

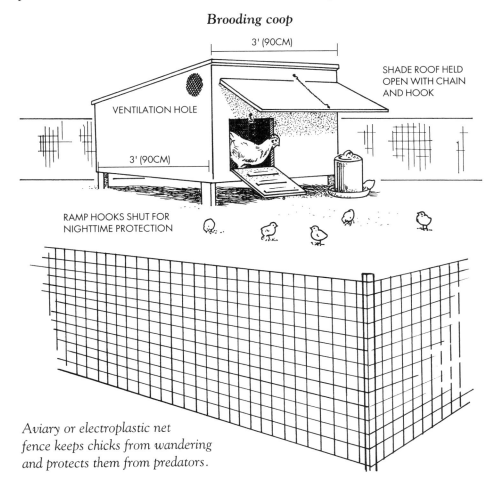

3' (90CM)

SHADE ROOF HELD
OPEN WITH CHAIN
AND HOOK

VENTILATION HOLE

3' (90CM)

RAMP HOOKS SHUT FOR
NIGHTTIME PROTECTION

*Aviary or electroplastic net
fence keeps chicks from wandering
and protects them from predators.*

while teaching her brood to scratch. You'll also need a water container the hen can't knock over. A 1-gallon plastic waterer is ideal. To protect chick-size feeding and watering stations from a too-exuberant hen, put up a barrier of 2-inch (5 cm) wire mesh the chicks can pop through to reach feed and water.

Hatching Schedule

A hen's decision to set is based on the changing season, as indicated when daylight hours begin increasing. If you hatch artificially, you'll do best not to stray too far from a hen's natural timing.

Breeder flocks are strongest and healthiest in spring, making spring chicks strongest and healthiest as well. Furthermore, chicks hatched in cool weather have time to develop immunities before the heat of summer causes germs to thrive. Chicks hatched in warm weather have little time to develop immunities against the proliferation of germs they are immediately exposed to.

Among show birds, large breeds mature in eight to ten months, bantams in six to seven months. To have young birds in prime condition for the fall round of shows, hatch large breeds in December, bantams no later than March.

Pullets hatched in spring begin laying by fall and continue to lay for about a year. Pullets hatched in winter will begin laying by mid-summer, but may molt and stop laying in the fall and won't start again until the following spring.

For general health and vigor, the best months to hatch in most areas are February and March. If you live in the far north where the weather stays cold long into spring, March and April are the best hatching months.

Artificial Incubation

An electric incubator is essential if you raise one of the breeds that doesn't tend toward broodiness, if you want chicks out of season, if you wish to hatch more chicks than your hens can handle, or if you simply don't trust your hens to pull off the job.

Electric incubators come in numerous styles and a wide range of prices. The greatest variety is in size. The smallest incubator, designed for classroom demonstrations, holds no more than four bantam eggs. At the other extreme, a large room-size commercial incubator hatches thousands of eggs at a time.

Affordable incubators for home use hold anywhere from 30 eggs to a few hundred. A larger, heavier cabinet unit has a windowed door at the front and is generally made of hard plastic or moisture-resistant wood. A smaller tabletop model has a removable cover and may be made of sheet metal (the so-called tin hen), hard plastic, or styrofoam. Styrofoam incubators are the least expensive but the most difficult to sanitize; my experience (and that of others) is that after a few hatches the success rate drops sharply.

In an effort to imitate the hatching conditions provided by a setting hen, incubators have four features:

◆ air flow
◆ temperature control
◆ humidity control
◆ turning device

Air Flow

An incubator needs good ventilation to bring in oxygen and to remove carbon dioxide and moisture generated by developing embryos. An accumulation of moisture on the incubator's viewing window during a hatch is a sure sign you need to improve ventilation by adjusting the vents.

Some incubators have vents with plugs you can remove as the hatch progresses, others have adjustable vents with sliding covers. A smaller incubator may rely on gravity to let warm, humid air escape through holes in the cover, which in turn causes fresh air to be drawn in through vents in the floor.

Better units have fans to circulate the air. Although a fan-ventilated (or forced-air) incubator costs more than a gravity-ventilated (still-air or natural-draft) incubator, it maintains a more uniform temperature throughout. To keep cool air from being blown onto the eggs, a forced-air cabinet model should have

Tabletop Incubator

Cabinet Incubator

an external switch that lets you turn off the fan when you're going to open the door for an extended period of time (for example, if you turn all the eggs manually).

Except for its blades, the fan should be sealed so it can't get gummed up by fluff and other hatching debris. Nevertheless, whenever you clean your incubator, give the fan a squirt of pressurized air or a lick with the vacuum cleaner.

Whether your incubator is forced-air or still-air, mark the placement of each egg the first few times you run it. You may find that eggs in certain spots won't hatch, so you can avoid putting eggs there in the future.

Temperature Control

High-tech incubators feature solid state temperature control. The temperature in most incubators is still regulated by a wafer, or thin disk, filled with ether. When the incubator heats up, the ether expands, causing the wafer to swell until it makes contact with a button switch that turns the heat off. As the incubator cools down, the ether contracts until the wafer loses contact with the switch, causing the heat to go on again.

Regulating the heat involves turning a control bolt to adjust the temperature at which the wafer makes contact with the switch. Most incubators have an indicator light that goes on when the heat is on and off when the heat is off. After a while, you'll develop a sixth sense that makes you feel uneasy if the light stays on or off for too long.

Sometimes a wafer "goes out," usually because it springs a leak and the ether escapes. In that case, the switch remains in the "on" position, causing the temperature to remain too high for too long. Unless you catch it right away, the eggs will cook. It pays to keep at least one spare wafer handy.

The wafer isn't the only thing that can cause the temperature to soar. The incubator could be too close to a heater, for example, or sunshine could fall on the incubator. Even a short-term rise in temperature does greater damage than a drop, which may occur if the incubator's cover or door was left ajar, hatching fluff has jammed the switch, a child or pet has pulled the plug, or the power has gone off.

Incubator Thermostat

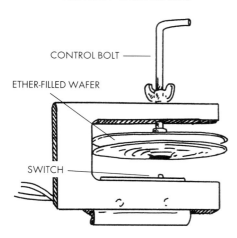

CONTROL BOLT

ETHER-FILLED WAFER

SWITCH

The wafer assembly consists of an ether-filled wafer that presses against a switch, and a control bolt to adjust the wafer's distance from the switch.

The more often you check the temperature, the more likely you are to catch and correct problems such as a pulled plug or a wafer gone bad. The first few times I used an incubator, I kept a checklist nearby on which I recorded the time and temperature every time I passed by. After a while checking got to be a habit, and I no longer needed the list as a reminder.

A higher priced cabinet incubator may be fitted with an alarm that goes off when the temperature drops or rises past a certain range. If you're handy with gadgets, you could easily rig up something similar. To protect an electronically controlled incubator from power fluctuations, install a surge suppressor.

To keep the temperature steady, locate the incubator in a room that has no drafts and has a fairly even temperature. The better the incubator is insulated, the less it will be affected by fluctuations in room temperature. Fluctuations are most often caused by sun coming through a window, the household heat or air conditioning being turned off during the night, or heavy storms that cause drastic changes in barometric pressure. Occasional minor fluctuations are normal, so once you regulate the temperature don't keep toying with it.

Power Outages

The more valuable your hatching eggs are, the more likely it is that the electricity will go out during incubation. If you have an uninterruptable power source (UPS) for your computer or other electronic equipment, consider disconnecting the usual equipment and using the UPS to power your incubator. If the outage continues beyond its capacity to keep your incubator running, or you don't have an UPS, open the incubator and let the eggs cool until the power goes back on.

Trying to keep the eggs warm is likely to cause abnormal embryo development. Furthermore, if you close the vents or wrap the incubator with blankets in your attempt to keep eggs warm, a greater danger than temperature loss is oxygen deprivation. Developing embryos use up oxygen rather rapidly, and the oxygen level may soon fall below that necessary to keep them alive.

As soon as the power goes back on, close the incubator and continue operating it as usual. The effect of the outage on your hatch will depend on how long the power was out and on how long the eggs had been incubated before the outage. A power failure of up to 12 hours may not significantly affect the hatch (except to delay it somewhat), especially if the outage occurred during early incubation, when cooled embryos naturally tend to go dormant. Embryos that are close to term generate enough heat to carry them through a short-term outage.

Make sure your thermometer is accurate by checking it against a second or even a third thermometer whose accuracy you are sure of. Position the thermometer according to the incubator manufacturer's instructions. Likely spots are in the viewing window of a cabinet model or at the height of the top of the eggs in a tabletop model. Since the air in a still-air incubator stratifies according to temperature, the thermometer *must* be at the same level as the eggs for good hatching results.

Even if the thermometer is accurate and is properly placed, you'll have a hard time regulating temperature if it's difficult to read. Most incubator thermometers have such small numbers that you can't read them without opening

Incubator Thermometers

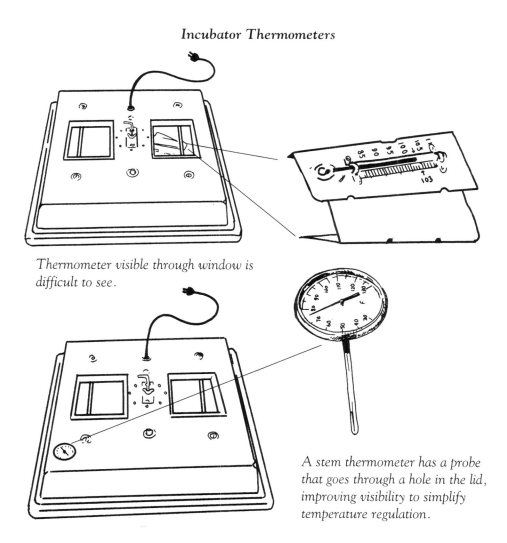

Thermometer visible through window is difficult to see.

A stem thermometer has a probe that goes through a hole in the lid, improving visibility to simplify temperature regulation.

the incubator, which of course causes the temperature to drop before you get a reading. For a cabinet incubator, I like to put a nice big thermometer in the window. For a tabletop incubator, I prefer a stem thermometer that sits on top and has a probe that goes through the cover.

Set the incubator to the temperature recommended by the manufacturer, which may vary from one model to the next. Typical operating temperatures are 99.5°F (37.5°C) for a forced-air incubator and 102°F (39°C) for a still-air incubator. Lethal temperatures are 103°F (39°C) in a forced-air incubator, 107°F (41°C) in a still-air incubator.

If the temperature runs slightly low, chicks will take longer than the normal 21 days to hatch. They will tend to be big and soft with unhealed navels, crooked toes, and thin legs. They may develop slowly or may never learn to eat and drink, and will therefore die. If the temperature runs slightly high, chicks hatch before the allotted 21 days. They tend to have splayed legs and can't walk properly, a problem that does not improve as the chicks grow. Such chicks should be culled.

Humidity Control

For a successful hatch, moisture must evaporate from the eggs at just the right rate. Too-rapid evaporation can inhibit the chicks' ability to get out of their shells at hatching time. Too-slow evaporation can lead to mushy chick disease (omphalitis), in which the yolk sac isn't completely absorbed so the navel can't heal properly. As a result, bacteria invade through the navel, causing chicks to die at hatching time and for up to two weeks afterwards.

Evaporation is regulated by the amount of moisture in the air — the more moisture-laden the air, the more slowly moisture evaporates from the eggs, and vice versa. To slow the rate of evaporation from eggs, every incubator has a water-holding device that releases moisture into the air.

The water-holding device might be a simple pan, a divided pan, or grooves molded into the incubator's floor. Humidity is regulated by the size of the pan or by the number of divisions or grooves you fill with water. Most devices have to be filled manually. Since cool water draws heat from the incubator, always add warm water.

To automatically add water as it evaporates, a few tabletop incubators come with a small reservoir similar to the upside-down bottle used to water pet rodents. Some styles of cabinet can be retrofitted with a 5-gallon reserve tank.

Humidity control may be fine-tuned by adjusting the incubator's vents. Opening vents decreases humidity by allowing moisture-laden air to escape. Closing vents increases humidity by trapping moisture-laden air within the incubator.

Low humidity tends to be a problem with small incubators and those that have to be opened to turn the eggs. If humidity drops inexplicably during a hatch, you may need to clean the water pan. Fluff released by newly hatched chicks will coat the surface of the water, preventing evaporation and causing the humidity to drop.

To measure the humidity in your incubator, you'll need a hygrometer. A hygrometer is little more than a thermometer with a piece of wet cotton cloth, called a "sock," wrapped around the bulb. A white cotton tennis-shoe lace makes a handy sock. Cut off a piece and tie it over the thermometer bulb with cotton thread. The reading, in wet-bulb degrees, tells you how much moisture is in the air based on how much moisture evaporates from the sock.

Some cabinet incubators have room for a permanent hygrometer and a reservoir that keeps the sock wet. You'll need spare socks to replace those that get crusty with mineral solids and lose their absorbency. If your tap water tends to be hard, each sock will last longer if you use distilled water.

You can make a temporary hygrometer by wrapping a piece of cheesecloth or gauze bandage around the bulb of a regular thermometer and dipping the cloth in warm water. Place the thermometer in your incubator where you can see it when the cover is on or the door is shut. After a few minutes the wet-bulb temperature will stabilize and you can take a humidity reading.

A forced-air incubator typically operates at wet-bulb 86°–88°F (30°–31°C) until the time of hatch, when humidity should be increased to wet-bulb 88°–91°F (30°–33°C). The typical hygrometer for a still-air incubator measures relative humidity, usually 60 to 62 percent during incubation, raised to 70 percent during the hatch.

A wet-bulb thermometer is more accurate in a forced-air incubator than in a still-air incubator. In place of a hygrometer, or in conjunction with one, a good indication of humidity is the changing air cell size inside the developing eggs. Moisture evaporating from an egg causes its contents to shrink, which in turn increases the size of its air cell. If your air cells are proportionately larger than the one shown in the accompanying sketch (as determined by candling the eggs), increase humidity; if they're smaller, decrease humidity.

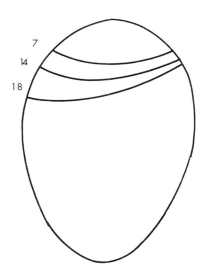

Relative air cell sizes on the 7th, 14th, and 18th days of incubation are indicated above.

As moisture from within an egg evaporates during incubation, the egg's weight decreases. Under proper humidity, an egg will lose some 10 to 12 percent of its weight by the eighteenth day of incubation. So another good way to gauge humidity is to weigh eggs at the time they are set and again on the eighteenth day. If the eggs weigh more than 90 percent of their starting weight, humidity is too high. If they weigh less than 88 percent of their starting weight, humidity is too low.

Temperature and Humidity

Optimum temperature and humidity are interrelated: as the temperature goes up, relative humidity must go down to maintain the same hatching rate.

No matter what combination of temperature and humidity is recommended by your incubator's manufacturer, you'll likely have to make minor adjustments to suit your specific conditions and the kind of eggs you hatch. Since small eggs evaporate more rapidly than large eggs, for example, they hatch better at a lower temperature and higher humidity than large eggs.

As you make adjustments, keep accurate records of the temperature and humidity for each hatch and of your success rate. After a few hatches, you'll home in on the optimum combination for your circumstances that will give you the best possible hatch.

Turning Device

By fidgeting in her nest and by adjusting eggs with her beak, a setting hen periodically turns each egg beneath her. Constantly adjusting her eggs may make the hen more comfortable or perhaps offers some psychological benefit. The hen can't possibly know that by turning each egg, she keeps its yolk centered within the white, and that if an egg isn't turned its yolk will eventually float away from the center and stick to the shell lining.

To imitate the hen's activities, some incubators have automatic turning devices that tilt the eggs from side to side at regular intervals, every hour or every four hours. In a cabinet model, whole shelves may tilt from one side to the other. In a tabletop incubator, tilting racks may be used to hold the eggs. The first time you use a tabletop, you may find that as the eggs turn, some of them hit the fan housing or the switching mechanism in the cover and get crushed. Mark the positions of those trouble spots in the racks so you can avoid using them in the future.

Some incubators, especially budget still-air units, don't have turning devices. Price may be an important consideration in deciding whether or not to get an incubator with a turning device, but your time may be of equal concern. Without a turner, you'll need to be on hand at least three times a day, every day, to turn the eggs yourself.

For manual turning if your incubator comes equipped with a rack, simply flip the eggs from one angle to another. If the eggs lie flat on a tray, to make sure you've properly turned each one place an "X" on one side and an "O" on the opposite side with a grease pencil or china marker. Some folks use a soft (number 2) pencil, but I stopped using pencil after I pierced a few shells. (Another reason I don't like to mark eggs with a pencil is that I don't always know ahead of time if I will hatch an egg in an incubator or under a hen, and penciled codes quickly rub off eggs hatched under a hen.)

Signs of improper turning are early embryo death due to a stuck yolk and full-term chicks that fail to pip, or break through the shell. Whether the eggs are turned automatically or manually, they aren't turned end for end, but side to side. Throughout incubation, the pointed end of the egg should never be oriented upward. Otherwise, according to research at the University of Georgia, fewer chicks will hatch, and the ones that do hatch will be of lower quality. Eggs do not need to be turned after the fourteenth day of incubation, and should not be turned during the last three days before the hatch, when chicks need time to get oriented and begin breaking out of their shells.

Candling Devices

To monitor the progress of each hatch so you can cull spoiled and nondeveloping eggs, you'll need a candling device. The best candler I ever had was a slide projector. When I held an incubated egg against the back-lit lens, I could clearly see everything inside, right down to the little beating heart.

When the projector died, I bought a hand-held candler that looks something like a small flashlight with a plug-in cord. To my surprise, I could only vaguely see what was going on inside an egg. One day when I couldn't find the candler, I discovered that the little flashlight on my key ring works better than the appliance designed for the job.

So before you go out and spend money on a candling device, look for something around the house that might work at least as well. All you need is light that comes through an opening smaller than the diameter of the eggs you want to candle. If all you can find is a flashlight, tape a 3-inch (7.5 cm) length of empty toilet-paper tube to the business end, so the only light you see comes through the tube.

Use your candling device in a dark room. Hold the egg at a slight angle, large end to the light and pointed end downward. Making sure your fingers don't block the light, turn the egg until either you see something or you're certain there's nothing to see.

Candling

White-shelled eggs are easier to candle than eggs with colored shells, which is why white eggs have become the industry standard. Similarly, plain-shelled eggs are easier to candle than eggs with spotted shells. A good way to gain practice candling is to hatch white-shelled eggs in your first setting. (Don't buy white eggs from the grocery, though; they're not fertile, but even if they were, they wouldn't hatch well due to refrigeration.)

When you examine an egg after one week of incubation, you will likely find one of three things:

- a webbing of vessels surrounding a dark spot — the embryo is developing properly
- a thin ring within the egg or around the short circumference — the embryo has died
- nothing (or a vague yolk shadow) — the embryo has died, or the egg was infertile

Assuming you remove infertile eggs and those with blood rings during the first candling, when you examine eggs after two weeks of incubation, you will likely find one of two things:

- a dark shadow except for the air cell at the large end (you may see movement against the air cell membrane) — the embryo is developing properly
- murky or muddled contents that move freely and/or a jagged-edged air cell — the embryo has died.

Statistically, embryos die at two peaks. The first peak occurs within a few days of the beginning of incubation. The second, larger peak occurs just before the hatch. Embryo deaths clustered during early incubation are most likely due to improper egg handling or storage. Embryo deaths during mid and late incubation may result from inadequate breeder flock diet. Failure of chicks to hatch is often caused by bacterial contamination — in other words, unclean eggs were placed in the incubator.

The multiplication of bacteria within a contaminated egg may cause the egg to rot and stink. Left in the incubator, a rotting egg may explode, spreading bacteria throughout the incubator and contaminating other eggs. Any time you open your incubator and smell an unpleasant odor, use your nose to ferret out the offending egg. It may or may not exude darkish fluid that beads on the shell, or (due to the pressure of gasses within the shell) it may be cracked and leaking.

Egg Culling

The first few times you candle eggs, break open the culls to verify your findings. (Tip: take the eggs outside in case the contents smell bad.) Even after you gain confidence in your abilities, continue breaking and examining culls to see what you can learn. You might learn something that will help you improve your future hatching success. Possible findings on candling include:

Blood spot or meat spot. A dark spot inside the egg is most likely a bit of blood or body tissue that accompanied the yolk when it was released in the hen's ovary. Cause: heredity or disease.

Tremulous air cell. The air cell has become detached and quivers or appears somewhere other than at the large end of the egg. As the cell moves around, the membrane will deteriorate and eventually rupture. Cause: egg was handled roughly, was stored at a high temperature, or was defective when laid.

Stuck yolk. As an egg ages and the white grows thin, the yolk floats toward the shell until eventually it sticks to the shell membrane. Cause: egg was left too long in one position during storage.

Seeping yolk. The yolk membrane has torn, letting yolk mix with the white. Cause: stuck yolk broke away from the shell membrane.

Addled egg or mixed rot. The contents of the egg are murky. Cause: bacterial contamination or late stages of seeping yolk.

Sour egg. A murky shadow surrounds an off-center swollen yolk floating in weak white (weak white lets the yolk sway when you move the egg during candling). If you break the egg open, the white looks greenish. The egg may or may not smell sour. Cause: *Pseudomonas* bacteria.

Black rot. Contents appear opaque or cloudy, except for the air cell. When you break the egg open, the contents are muddy brown and smell putrid. Cause: bacteria in the *Proteus* group.

Assuming you incubate only properly stored clean eggs, you should find few culls, and the majority of your fertiles should hatch. The average hatching rate for artificial incubation is about 85 percent of all fertile eggs.

A wonderful book that shows photographically all that goes on inside a properly developing egg is A *Chick Hatches* by Joanna Cole and Jerome Wexler. This fine book is now out of print, but it's well worth looking for at your local library, borrowing through interlibrary loan, or searching for in used bookstores.

Improve Your Hatching Rate

To optimize incubator space, incubate a full setting for 15 hours, then candle and cull infertiles and those with germinal discs that are smaller than the majority. Put the remaining eggs back into storage and repeat the process with a second setting. Combine the best of both settings and incubate as usual. A short dormancy period will not affect hatchability. Instead, you'll improve your overall hatching rate.

Hatchery Sanitation

Illness in freshly hatched chicks comes from one of two sources: poor sanitation in the incubator and/or around the brooder, and disease in the breeder flock that's transmitted through the eggs. Some egg-transmitted diseases enter an egg as it's formed within an infected hen. Others are caused by organisms that get on the shell as the egg is laid, when it lands in a contaminated nest, or when it is improperly washed.

Common diseases that can be transmitted through hatching eggs include:

◆ chronic respiratory disease
◆ lymphoid leukosis
◆ mushy chick disease (omphalitis)
◆ paratyphoid
◆ pullorum
◆ typhoid
◆ viral arthritis

These diseases spread from infected chicks to healthy chicks in the incubator (often inhaled in fluff) or in the brooder (usually through ingested droppings in feed or water). Of these, the most common is omphalitis. It has a number of causes including incubation of dirty eggs, operation of an unsanitary incubator, and improper temperature or humidity during incubation.

Good incubator sanitation not only improves hatching success and gives chicks a healthy start in life but also helps break the natural disease cycle in a flock. Because hatching itself is a major source of contamination, clean your incubator after each hatch. Vacuum up loose down, sponge out hatching debris, and scrub the incubator with detergent and hot water. When the incubator looks thoroughly clean, apply a disinfectant such as Germex or chlorine bleach (¼ cup Clorox per gallon of hot water, or 30ml/l). If possible, dry the incubator in the sun.

Continuous Hatching

The best sanitary measure is to incubate one setting of eggs, then clean and disinfect the incubator before making another setting. I, for instance, fill my tabletop incubator with eggs from my New Hampshires, hatch them, clean out the mess, then make a second setting.

In the days when I raised several show breeds and had only a trio of each, I used a cabinet incubator with four trays. I filled one tray each week with eggs from all breeds and used the fourth tray for the current hatch. This practice of continuous hatching is a good source of incubator contamination.

> **Caution**
>
> Do not feed hens the shells from incubated eggs — they're loaded with bacteria and can spread disease.

To minimize contamination, I moved the hatch to the lowest tray where no lower eggs could get covered with hatching debris. After each hatch, I removed the hatching tray, and scrubbed and disinfected it. While I was at it, I cleaned and disinfected the water pan as well. Then I moved all the trays in the incubator down one slot and placed the cleaned tray at the top, filled with fresh eggs. This rotation ensured that each tray was cleaned every three weeks.

Whether you use a cabinet incubator or a tabletop model, an excellent sanitary measure is to keep a separate, smaller incubator as a hatcher. Ideally it will be a forced-air incubator, but it doesn't need a turning device. When your hatcher is properly regulated, move the about-to-hatch eggs from the incubator to the hatcher. After the hatch is complete, scrub and disinfect the hatcher in preparation for the next hatch.

The Hatch

A normal hatch is complete within 24 hours of the first pip. A dragged-out hatch, or one that occurs earlier or later than you expected, is likely caused by improper incubation temperature or humidity. Another cause for so-called draggy hatch is combining eggs of various sizes in one setting. Larger eggs, as well as eggs that have been stored longer, tend to take longer to hatch than smaller eggs. If you combine eggs from bantams and large fowl, or from light and heavy breeds, the bantam eggs or those from the lighter breeds will usually hatch first.

When the hatch is short and quick, move the chicks to a brooder when most of them (95 percent) have dried and fluffed out. Moving chicks while

All chicks should hatch within 24 hours of the first pip.

they're still wet can cause them to chill. If the hatch is draggy, remove the dried chicks every 6 to 8 hours so the incubator's air flow won't cause them to dehydrate. Work quickly, as opening the incubator will cause the temperature and humidity to drop, reducing the percentage of remaining eggs that will hatch.

Pedigree Baskets

If you hatch eggs from different matings in the same setting, keep track of which chicks came from which mating by using pedigree baskets. These might be nothing more than upside-down pint-size plastic fruit baskets set over groups of eggs. You can also fashion pedigree baskets from ¼-inch or ½-inch (6.25–12.5 mm) hardware cloth, bent to form a box 4-inches to 6-inches (10–15 cm) square and about 2 inches (5 cm) high. Cut a cover to fit, and hinge one edge with wire. A couple of paper clips along the opposite edge will hold the lid down. Alternatively, you might make pedigree sacks from fine-mesh netting, such as the plastic mesh bags onions sometimes come in.

On the eighteenth day of incubation, group the eggs into their baskets or sacks and paper clip a record of what's inside to each. After the chicks have hatched, remove one group at a time and place an identifying mark on each chick in the group.

Chick Identification

Using pedigree baskets makes sense only if you have a way to keep track of the chicks after they hatch. One way, of course, is to raise each group separately from the others. But that can be impractical if you hatch chicks from several different matings. Chicks from different matings can be marked for identification in one of three ways.

Toe punching involves removing flaps of skin between the forward three toes on one or both of a chick's feet. The patterns formed by removing or not removing a web form 16 possible combinations, as shown in the illustration. Each pattern is assigned a number from 1 to 16, and all chicks from one mating are punched with the same pattern. The pattern for 16 is the least desirable, since no webs are punched — you'll always wonder if this chicken is *really* from #16 mating or if it once had punched webs that later healed back together.

The toe punching tool, available from nearly any poultry supply catalog, is about the size of a pair of fingernail clippers and functions much like a paper hole punch. As you remove each chick from its pedigree basket, hold it gently but securely in one hand with one foot extended. Carefully position the punch over the web to be removed. With one firm stroke, punch away the front part of the web. Don't punch a hole through the web, or the web may eventually grow back together. A well-placed web punch is permanent.

Wing banding involves applying a numbered metal band to the wing web — the wide flap of skin between the bird's wing and its body — as soon as the chick fluffs out. Although wing bands are more often used for large chickens than for bantams, there's no reason they can't be applied to either. The advantage to wing bands is that they give you more numbers than are available from toe

Toe Punch

Insert punch between two front toes and press hard to remove the web.

Toe Punch Patterns

LEFT FOOT		RIGHT FOOT
\o| /	1	\ | /
\ |o/	2	\ | /
\o|o/	3	\ | /
\ | /	4	\o| /
\o| /	5	\o| /
\ |o/	6	\o| /
\o|o/	7	\o| /
\ | /	8	\ |o/
\o| /	9	\ |o/
\ |o/	10	\ |o/
\o|o/	11	\ |o/
\ | /	12	\o|o/
\o| /	13	\o|o/
\ |o/	14	\o|o/
\o|o/	15	\o|o/
\ | /	16	\ | /

punching. Assigning each chick its own number, rather than a toe-punched mating code, lets you accurately track individual birds throughout their lives.

Two kinds of wing bands are available through poultry supply catalogs. One is inserted in place with a pair of wing-banding pliers. The other is inserted by hand (giving you better control over placement), then sealed with wing-banding pliers. To avoid tearing the wing web, apply the band after chicks have fluffed out and have begun to toughen. Check week-old chicks and adjust any band that has slipped over a tiny wing.

Except in tightly feathered breeds, a wing band is clearly visible only on close examination. It provides a permanent means of identification, unless you deliberately choose to remove it.

Applying a Wing Band

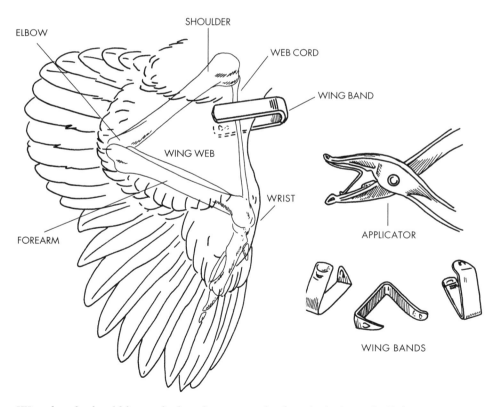

Wing bands should be applied to the wing web after chicks have fluffed out. Two kinds are available: those that must be applied with an applicator and those that can be applied by hand.

Leg Banding is the least desirable means of identifying chicks, since the bands have to be changed several times as little legs grow. Apply bands as soon as you remove chicks from the hatcher, taking care not to break delicate legs.

For chicks of the large breeds start with #4 bands; for bantams you'll need #3 or maybe #2. As the birds' legs grow and the bands become so tight you can no longer easily slide them up and down, switch to a larger size. Never leave a band on so long it begins to bind or you'll end up with a lame bird.

You have the option of using either plastic spiral rings to identify matings by color code or numbered bands to identify individual birds. Spiral rings are thinner

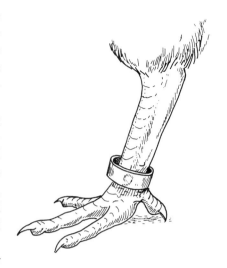

Leg bands help you identify individual chickens.

and therefore easier to apply on little legs than the wider numbered bands. There's no reason not to start with spirals and switch to numbered bands later on. Both kinds tend to break, making leg banding less foolproof than the other two identification methods. Permanently sealed leg bands can't be applied until birds have grown to full size.

Recordkeeping

Hatching records are essential not only for fine-tuning your incubation procedures but also for helping you pinpoint breeder-flock management problems that affect the hatch. The more detailed your records are, the easier you'll be able to spot patterns that show room for improvement. Leave space at the bottom to note such things as the dates and times of power outages or changes in your incubation procedure you might wish to try in the future.

If you mark your chicks with individual identification, set up a second sheet with a line for each chick and space for any data you find important, such as the chick's down color and (later, when it becomes apparent) its gender. Under "notes" you might keep track of such things as deaths, reason for culling (if any), or changes in plumage color.

Hatching Record

Breed/pen: _____ Cock: _____

Date set: _____ Time set: _____

Date candled: _____ Results: _____

Date candled: _____ Results: _____

Stop turning on: _____

Hatch should start on: _____

Hatch started on: _____ Time:_____ am/pm

Hatch ended on: _____ Time:_____ am/pm

Toe punch code: _____ Band color code: _____

RESULTS:

A. Number of eggs set _____   G. Culled, too weak _____

B. Not fertile _____   H. Culled, deformed _____

C. Died 1st week _____   I. Other _____

D. Died 2nd week _____   J. Total lost _____

E. Full-term, failed to pip _____   K. Total hatched (line A minus line J) _____

F. Pipped, failed to hatch _____   L. Percent of fertiles hatched

(line K divided by [line A minus line B]) _____%

NOTES

Chick Record

Number	Sire	Dam	Hatch	Date	Sex	Color	Notes

Chapter 10

Chick Care

NEWLY HATCHED CHICKS aren't entirely helpless, but until they grow a full complement of feathers, you'll need to keep them warm and dry, and protect them from dogs, cats, and other animals. Like any other babies, they must also be kept clean and well fed.

Brooders

A brooder is a place where chicks are temporarily raised until they have enough feathers to keep themselves warm. A brooder should provide:

- adequate space
- protection from predators
- protection from moisture
- a reliable heat source
- freedom from drafts
- good ventilation

Brooders come in many sizes and styles. The style that's best for you will depend on how often you plan to raise chicks and how many you wish to brood at once. Whatever style you choose, locate your brooder where you won't be bothered by the dust chicks invariably stir up due to their constant activity.

A *battery* consists of a series of tiers stacked one on top of the other. Each level has built-in feeders and waterers along the sides and front, a wire mesh floor, and a removable tray to catch waste so it doesn't fall on the chicks below. To see a battery at work, visit a farm store that sells chicks. Farm stores favor battery brooders because a battery lets you keep lots of chicks in a small space, and the tiers offer a convenient way to separate different breeds or species.

The number of chicks you can brood in a battery depends on the dimensions of each tier and the number of tiers. Some batteries hold a set number of tiers; others come as individual stackable units. Each level has its own electric heater coil, which allows you to heat only as many tiers as you are using at the time.

Common practice is to keep chicks in a battery for 7 to 10 days, then move them to roomier quarters. You *can* battery-rear broilers to 3 pounds, but they'll likely get breast blisters unless you cover the wire mesh with plastic mats, which are messy to clean. In any case, move birds from a battery before they get so big their heads rub against the ceiling or they can't get their beaks through the wire feeder guards to eat.

A battery brooder lets you keep lots of chickens in a small space.

New batteries are available through farm stores and poultry supply catalogs. Since they're fairly expensive, you might do as I did and keep an eye out for a good deal on used ones. To operate a battery, you'll need a draft-free but well-ventilated garage or outbuilding where the temperature remains constant.

A *hover* is a sort of metal umbrella that hangs from the ceiling so it "hovers" over the chicks. Chicks can huddle beneath the hover to be warmed by its electric or gas heater, or they can move away from the heat to eat and drink. The hover may be round or rectangular, and may have curtains hanging around the edges to keep in heat and keep out drafts.

Hovers — sold by many farm stores and other poultry suppliers — offer an economical way to brood 100 chicks or more. The chicks are housed on a litter-covered floor, either in a small predator-proof outbuilding or in a draft-free stall or corner of a larger building. Feeders and waterers are spaced around the outer edges of the hover where chicks can easily find them.

A *heat lamp* is basically an inexpensive hover. For 25 to 100 chicks, one 250-watt infrared heat lamp is sufficient. If your brooding area is large enough to handle the extra heat, you're better off using two lamps in case one burns out when you're not around.

Hang each lamp by an adjustable chain, starting 18 inches (45cm) above the floor. An infrared bulb gets quite hot, so use porcelain rather than plastic sockets, since the latter will melt. To avoid a fire hazard due to the lamp

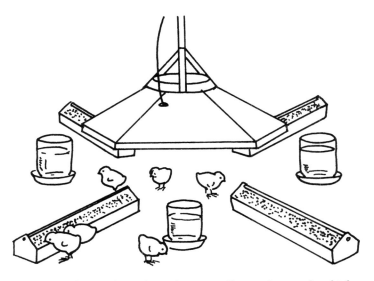

A hover is a heater that hangs from the ceiling so it "hovers" over the chicks to keep them warm.

coming into contact with paper, litter, wood, or other flammables, use only fixtures with wire guards.

To confine the chicks close to the heat, you might house them in a wooden or cardboard box or in an old stock tank. Secure a piece of wire mesh over the top to keep out cats and other chick eaters, and lay a piece of cardboard or plywood across part of the top, if necessary, to keep out drafts.

Infrared lamps with either red or clear bulbs are available at farm stores, electrical supply outlets, and some hardware stores. A red lamp is more expensive than a white lamp but won't burn out as quickly, and the red glow discourages picking.

An incandescent light bulb provides the least expensive heat source for batches of 25 to 50 chicks. Screw

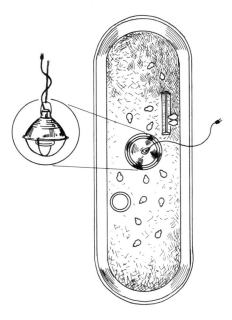

An infrared heat lamp with an aluminum reflector and a porcelain socket, hung over a stock tank lined with litter, makes a dandy brooder.

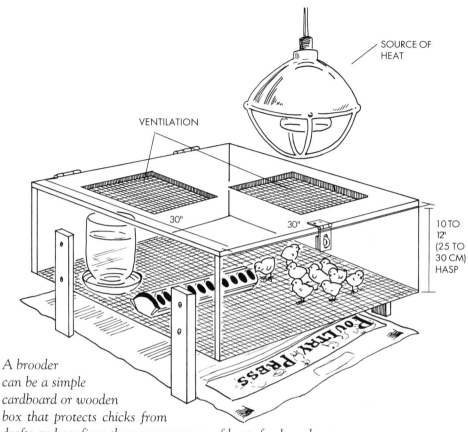

SOURCE OF
HEAT

VENTILATION

30"

30"

10 TO
12"
(25 TO
30 CM)
HASP

*A brooder
can be a simple
cardboard or wooden
box that protects chicks from
drafts and confines them near sources of heat, food, and water.*

the bulb into a fixture with a reflector and hang it over confined chicks, or use the bulb to create a so-called floor hover by nailing a large metal can to the floor and hanging the bulb inside the can, so chicks can gather around it like old-timers around a potbelly stove.

Brooder Guard

If you brood chicks on the floor, you'll need some way to keep them from wandering far from the heat source and getting chilled. A brooder guard surrounds the heat source to not only confine the chicks but also reduce drafts and eliminate corners where chicks tend to pile up and smother. Some farm stores carry ready-made brooder guards, but you can easily make your own by cutting up a large cardboard box.

A brooder guard is nothing more than a 12-inch (30 cm) high circular fence of corrugated cardboard. Place the fence 2 to 3 feet (60–90 cm) from the heat source, and fasten the ends together with clothespins or spring-type paper clips. In about 7 to 10 days, after the chicks have become fully familiar with the locations of heat, feed, and water, either remove the brooder guard or, if it's still needed to reduce drafts, expand it to give chicks room to grow.

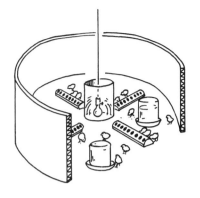

Floor hover provides heat; a brooder guard eliminates drafts.

Heat

A chick's body has little by way of temperature control until the bird is about 20 days of age, when down starts giving way to feathers. Chicks tend to feather out more quickly in cooler weather, but if air temperatures are quite low, they may need auxiliary heat longer than chicks brooded in warmer weather. Chicks hatched in winter or early spring require brooder warmth for 8 to 12 weeks. Chicks hatched in late spring or early summer need heat for 6 to 8 weeks.

Start the brooding temperature at approximately 95°F (35°C) and reduce it approximately 5°F (3°C) each week until the brooder temperature is the same as room temperature. Within the chicks' comfort zone, the more quickly you reduce the heat level, the more quickly the chicks will feather out.

Batteries and hovers operate by adjustable thermostat. An infrared lamp must be raised or lowered to adjust the temperature. A general rule is to raise the lamp about 3 inches (7.5 cm) each week. Be especially watchful with chicks confined in a small space, since infrared can get quite hot and you don't want the chicks to be cooked alive. As they get older and require less heat, give chicks more room so they can move away from the heat, or switch from an infrared lamp to an incandescent light bulb. Heat put out by a light bulb can be adjusted two ways: by raising or lowering the bulb and by decreasing or increasing the wattage. Start with 100- or 60-watt bulbs, depending on the size of the brooder and number of chicks.

Theoretically, brooder temperature is measured 2 inches (5 cm) above the floor (and at the outer edge of a hover), but you shouldn't need a thermometer. Just watch the chicks and adjust the temperature according to their body language.

Chicks that aren't warm enough — due either to insufficient heat or to draftiness — will crowd near the heat source and peep loudly, and may develop pasting or outright diarrhea. In an effort to get warm while they sleep, the

Body Language and Brooder Heat

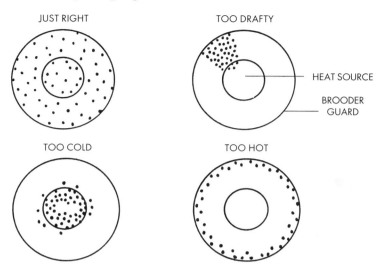

Watch where chicks stand in relation to one another to determine whether the temperature needs adjustment. (Dots represent chicks.)

chicks will pile up and smother each other. Smothering usually occurs at night when the temperature drops.

Chicks that are too warm move away from the heat, spend less time eating, and as a result grow more slowly. They'll pant and tend to crowd to the brooder's outer edges, perhaps smothering one another. If the heat is enough to raise their body temperatures over 117°F (47°C), chicks will die.

Chicks that are warm and cozy wander freely throughout the brooding area, emit musical sounds of contentment, and sleep side-by-side to create the appearance of a plush down carpet.

Light

Light affects the growth rate of chicks, so never keep them in the dark. To help them find feed and water, light the brooder continuously for the first 48 hours. Since chicks tend to move toward light, most commercial hovers have a so-called attraction light. For other types of brooder, one 25-watt bulb will adequately light 10 square feet (1 sq m). After the first two days, you can turn lights off if the brooding house gets natural daylight. Windows on the south side furnish the best sunlight.

Even if the light is also your source of heat, turn it off for ½ hour during each 24-hour period (preferably not during the coolest hours) so the chicks will learn not to panic later when the lights go out at night or if the electricity fails.

Putting the light on a timer will save you the trouble of remembering to turn it off and on each day.

Litter

Brooding chicks on wire mesh flooring has both advantages and disadvantages. The biggest advantage is ease of cleaning. The biggest disadvantage is the lack of gradual exposure that allows floor-reared chicks to develop immunity to coccidiosis, a major chickhood disease. Unless the birds spend the rest of their lives in wire cages, those brooded on wire are likely to suffer an outbreak of coccidiosis when moved to open housing.

Brooding chicks on litter from the start gives them the gradual exposure to coccidia they need to develop immunity. Litter also helps keep them dry, insulates the floor for added warmth, and absorbs droppings. Ideal kinds of litter are peat moss, wood shavings (pine *not* hardwood), crushed corncobs, crushed cane, and vermiculite. Sand and chopped straw are less desirable but are okay if that's all you can get. Avoid whole straw, since it mats down and chicks have trouble walking on it.

If you floor brood chicks on dirt, first cover the dirt with well-anchored ½-inch (12.5 mm) mesh aviary netting to keep out burrowing predators. Then spread at least 3 inches (7.5 cm) of litter.

For the first two days, until newly hatched chicks are eating well, keep them from pecking in the litter by covering it with paper toweling or other rough material. Do not use newspaper or other smooth paper, which is hard for little guys to walk on and can cause leg injuries. Change the paper toweling as often as necessary to keep it clean.

Once the toweling is removed, stir up the litter every day to keep it from packing down, and add a little fresh litter as necessary to keep it fluffy and absorbent. Replace any moist litter that develops around waterers, since damp litter quickly turns moldy and can cause aspergillosis, known as "brooder pneumonia." Many brooder diseases are easily prevented by proper litter management.

Space

In a battery, allow at least 10 square inches (65 sq cm) of space per chick up to 2 weeks of age, then increase the available space per chick by at least 10 square inches (65 sq cm) every 2 weeks. Under a hover or heat lamp, provide ½ square foot (450 sq cm) per chick up to 4 weeks of age, then 1 square foot (900 sq cm) to the age of 8 weeks. More isn't necessarily better — if you give chicks too much space during cold weather, they'll have trouble staying warm.

Brooding Space

Age	Square Feet	Square Cm
0–4 weeks	½	450
4–8 weeks	1	900
8–12 weeks	2	1800
12+ weeks,		
light breeds	2½–3	2250–2700
heavy breeds	3–3½	2700–3150

Grower Houses

As chicks grow, they need increasingly less heat and more space. By the time they are fully feathered at about 6 weeks of age, they need at least 1 square foot (900 sq cm) of space each. How often you move your chicks to new facilities depends on your management style. At our place we use a four-step system for our dual-purpose flock. We start chicks in a homemade wooden brooder set up in the mudroom (if the weather is nasty) or in the barn.

After about two weeks we move the chicks to a grower house that gives them more room to move around in and has an outdoor porch where they can get sunlight and fresh air. We confine them to the indoor portion of the grower house until they're big enough and the weather is warm enough for them to spend time on the porch.

A key feature to our grower house is the ability of its inhabitants to see what's going on in the yard around them. Chicks that see you coming are less likely to become wild than those caught by surprise every time you show up to tend them. If your housing design does not allow your chicks to see when you approach, the polite thing to do is announce your presence by talking or singing whenever you go near.

Another feature of our grower house is beginner's perches. When light breeds reach about 4 weeks of age and heavy breeds about 6 weeks, they're ready to roost on low perches. Don't provide roosts for broilers or they'll develop breast blisters and crooked keels. For others, nighttime roosting is a natural and healthy habit, and hopping on and off perches during the day is good entertainment. Allow 4 inches (10 cm) of roosting space per chick. Provide low and high perches, or start the perches close to the floor and move them up as the majority of birds learn to use them.

Grower House Roost

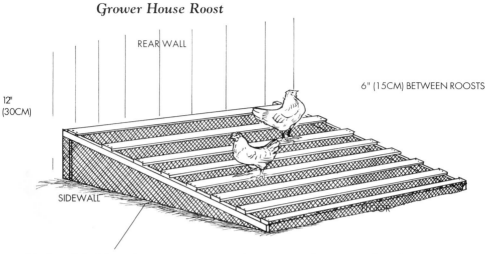

Tack ½" (125 mm) aviary netting under roosts and across ends to keep chicks from picking in droppings.

When our chicks are nearly feathered out and are at the age when we can tell the sexes apart, we move the pullets to a portable range house and leave the cockerels in the grower house until they reach fryer size. Separating the sexes helps both parties grow more steadily: pullets spend more time eating than running away from cockerels, and cockerels spend more time eating than running after pullets. When our pullets reach laying age, our old layers go into the freezer as stewing hens, the layer house is thoroughly cleaned, and the pullets are moved in as replacement layers.

This system lets us raise young birds away from the older flock and its potential for transmitting disease, and provides us with plenty of clean meat and year-round fresh eggs. We're lucky to have enough land to set up a multi-stage growing system.

There's no reason, however, why chicks can't be brooded from the start in what will be their permanent housing, provided it can be closed up tightly enough to protect young birds from drafts and predators, and has windows you can open to let in fresh air on warm days. If the housing has been used for chickens in the past, it needs to be thoroughly cleaned, swept free of dust and cobwebs, and washed down — first with warm water and detergent, then with warm water and chlorine bleach or other poultry-safe disinfectant.

Many people these days prefer to raise chicks on pasture. Although the concept has been getting a lot of attention lately, before about 1950 most chicks were reared on range. A portable range shelter on skids is especially attractive for raising a batch of meat birds, since they don't require permanent year-round housing. Even when birds must later be moved to winter housing, range rearing offers sunshine, fresh air, and green feed that combine to keep them healthy. For proper growth, don't expect pasture to replace more than 15 percent of the chicks' usual ration.

Warm weather permitting, chicks can be moved from the brooder to a range shelter any time after they reach 6 weeks of age. For the first few days, confine them to the shelter or to a small yard enclosed with poultry netting so they can get oriented before you give them freedom to roam. If it rains, especially a long hard rain at night, go out and check your birds to make sure they aren't huddled outside or sleeping in a puddle inside the shelter.

Even if you don't have pasture or range, when the chicks are at least two weeks old and the weather is warm and sunny, you can put them in a pen on the lawn for a few hours each day, provided the lawn has not been sprayed with toxins. They'll need shade and water, a ½-inch (12.5 mm) mesh wire guard around the perimeter so they can't wander away, and wire mesh over the top to keep out hawks and cats. For good sanitation and to provide fresh forage, move the pen to a new spot each day.

If you have a garden, you can put the pen there. Ideally, it should be on sand or sandy soil that's spaded under after each move. If this strikes you as a routine of fertilizing and rototilling, you might at this point turn to page 251, "Chickens and Gardens."

Watering Chicks

Chicks can go without water for 48 hours after they hatch, but the sooner they drink the less stressed they'll be and the better they'll grow. A chick's body needs water for all life processes including digestion, metabolism, and respiration. Water helps regulate body temperature by taking up and giving off heat, and it carries away body wastes. A chick that loses 10 percent of its body water through dehydration and excretion will experience serious physical disorders. If the loss reaches 20 percent, the chick will die.

As you place chicks into the brooder, dip each one's beak into the water fount and make sure it swallows before releasing the bird. Some chicks start life quite active and curious, while others initially aren't very adventuresome, so for the first day or two keep waterers fairly close to the heat source where all chicks can easily find them.

Various health boosters may be added to drinking water to give chicks the best start in life. If the chicks have been shipped, a vitamin solution for the first three weeks will help them overcome shipping stress. Even if they haven't been shipped, a vitamin solution for at least one week will help them cope with the stress caused by the transformation from embryo to chick.

To give droopy chicks a spurt of energy, dissolve sugar in their drinking water at the rate of 1½ cups per gallon, or about ⅓ heaping cup per quart (95 ml sugar per liter). To suppress *Salmonella enteritidis*, mix up a 5 percent skim milk solution by adding enough water to ¾ cup skim milk to total 1 gallon, or about 3 tablespoons skim milk per quart (50 ml skim milk per liter). To avoid inducing diarrhea, take it easy on the milk and don't use whole milk.

Chick Water Requirements

A chick's need for water increases as the bird grows. To determine approximately how much water a batch of chicks needs, divide the chicks' age in weeks by 2 and you'll know how many gallons you should provide per 100 chicks per day. For example, 4-week-old chicks need about 2 gallons of water per day per 100 chicks, or 1 gallon per 50 chicks. (In metric, multiply 2 times the age in weeks to determine how many liters of water 100 chicks need per day; 100 4-week-old chicks need about 8 liters of water per day.)

Chick Waterers

To avoid dehydration, be sure chicks have access to fresh, clean water at all times. The easiest way to provide water to newly hatched chicks is to use a 1-quart canning or mayonnaise jar fitted with a metal or plastic watering base, available from most feed stores and poultry supply catalogs. A plastic base will crack over time, and a metal one will break away from the portion that screws onto the jar, so keep a few extra bases on hand.

If you can't find a chick watering base, or yours springs a leak, make an emergency waterer from an empty can and an aluminum pie tin that's 2 inches (5 cm) larger in diameter than the can. Bore two small holes on opposite sides of the can, ¾-inch (1.9 cm) from the open end. Fill the can with water, invert the pie tin on top, and turn the assembly upside down so the pan becomes a water-filled basin the chicks can drink from.

Don't be tempted to cut corners by supplying water in an open dish or saucer. Chicks will walk in it, tracking litter and droppings that cause disease. They'll tend to get wet and chilled, and the stress will open the way to disease. Some chicks may drown.

Within a week or so, the chicks will outgrow the 1-quart watering jars. You'll know the time has come when you find yourself filling founts more and more often to keep your chicks in water at all times. Another sign is finding chicks perched on top, a habit that lets droppings fall into the water. You'll know the time for changing founts has long since passed when you find the waterer overturned and the water spilled due to boisterous play in the brooder.

When it's time to change waterers, you have several choices. Regardless of the design, a good waterer has these features:

- It is the correct size for the flock's size and age — chicks should neither use up the available water quickly nor be able to tip the fount over.
- The basin is the correct height — a chick drinks more and spills less when the water level is between its eye and the height of its back.
- Chicks can't roost over or step into the water — droppings plus drinking water equal a sure formula for disease.
- The fount is easy to clean — a fount that's hard to clean won't be sanitized as often as it should be.
- The fount does not leak — leaky waterers not only run out fast but also create damp conditions that foster disease.

Automatic waterers are a great time-saver, are the most sanitary of all designs, and ensure that chicks won't run out of water when no one is around to refill founts. Another style is trough-type waterers. You'll need 20 lineal inches (50 cm) of trough per 100 chicks up to 2 weeks old, and 30 lineal inches (75 cm) thereafter. A trough that allows chicks to drink from either side offers twice the watering space of a one-sided trough.

A third alternative is to continue using the same style of fount, only in larger versions. Depending on the number of chicks you're brooding and the amount of space they have, you might first switch from quart jars to 1-gallon plastic founts, then move up to 3-gallon or 5-gallon metal founts. Any time you use indoor metal waterers connected directly to water lines, if chicks refuse to drink, stick a finger into the water. If you feel a buzz, a bad electrical connection such as poor grounding is causing chicks to get a shock each time they try to drink. Fix the problem if you can or call in an electrician.

Clean waterers daily with warm water and chlorine bleach. Initially place founts no more than 24 inches (60 cm) from the heat source. Later, as you move the chicks to expanded housing, make sure they never have to travel more than 10 feet (300 cm) to find a drinker. Whenever you change waterers, leave the old ones in place for a few days until the chicks get used to the new ones. This stress-reducing measure applies whether or not the chicks are moved at the same time the waterers are changed.

Chicks should have access to fresh, clean water at all times.

To keep chicks from picking in damp litter around the waterers, after the first week place each fount over a platform created from a 2-foot square (100 sq cm) frame made of 1 x 2-inch (2.5 x 5 cm) boards, covered with ½-inch (12.5 mm) mesh hardware cloth. Drips will fall below the mesh where chicks can't walk or peck.

Feeding Chicks

Feed chicks within three to five hours after they had their first drink. They'll experience less stress if they eat on the first day after they hatch, even though they can survive for another day or two without eating. Yolk reserves provide nutrients that, in nature, allow early-hatched chicks to remain in the nest until all the stragglers have hatched. In modern times, these same yolk reserves sustain newly hatched chicks while they're being shipped by mail.

The best ration for chicks is chick starter, which is higher in protein and lower in energy than rations designed for older birds. If you run out of starter, or you forget to pick some up and suddenly you have chicks to feed, make an emergency ration by running a little uncooked oatmeal through the blender and combining it 50/50 with cornmeal. Don't use your homemade ration any longer than necessary, since it lacks the nutritional balance needed for proper growth and disease prevention. And *do not feed lay ration to chicks*, even as an emergency feed — its high calcium content can seriously damage their kidneys.

Some brands of starter are medicated; some are not. Whether or not you need medicated starter depends in large part on your management style. Use medicated starter if:

♦ you brood chicks in warm, humid weather
♦ you brood large quantities of chicks at a time
♦ you brood one batch of chicks after another
♦ your sanitation isn't up to snuff

You shouldn't need medicated starter if you brood in late winter or early spring (before warm weather allows coccidia and other pathogens to flourish), you brood on a noncommercial scale, you keep your chicks on dry litter, and your chicks have only fresh clean water to drink.

In our area, as in many parts of the country, farm stores carry only one all-purpose starter or starter-grower ration. In areas where chickens are big business, farm stores may offer grower and/or finisher rations for broilers and grower and/or developer rations for layers and breeders. Switch from one ration to another as indicated on the labels, gradually making the switch by combining the old ration with greater and greater amounts of the new ration to avoid problems related to digestive upset. If you can't find grower, finisher, or developer in your area, continue using starter until broilers reach butchering age and layers/breeders are ready for the switch to lay rations.

As a general rule, you can expect each chick to eat approximately 10 pounds (4.5 kg) of starter ration during its first 10 weeks of life.

Chick Feeders

To help freshly hatched chicks find feed and learn to peck, sprinkle a little starter ration on a paper towel, a paper plate, a piece of cardboard, or heavy paper. As soon as most chicks are pecking freely, remove the paper before it begins to hold moisture that attracts mold.

For the remainder of the first week, put the starter in a shallow lid or tray. A shoe box lid or a tray of similar size is just right. When the chicks start vigorously scratching, switch them to a regular chick feeder, available from farm stores and through poultry supply catalogs.

Like chick waterers, chick feeders come in several styles. A good feeder has these features:

- ◆ prevents chicks from roosting over or scratching in feed
- ◆ has a lip to prevent billing out
- ◆ can be raised to the height of the birds' backs as they grow
- ◆ is easy to clean.

One style of chick feeder is a base, similar to a waterer base, that screws onto a feed-filled quart jar, and has little openings through which the chicks can peck. Another style is a trough, which may have a lid with individual openings chicks can peck into, or either a reel to keep chicks from perching over the feed or a grill to keep them from scratching in it.

If you use a feeder with individual openings, allow one slot per chick. If you use an open trough, allow 1 lineal inch (2.5 cm) per chick to 3 weeks of age, 2 lineal inches (5 cm) to 6 weeks of age, and 3 lineal inches (7.5 cm) to 12 weeks. Count both sides if chicks can eat from both sides.

Like grown-up chickens, chicks waste feed through billing out — the habit of scratching out feed with their beaks. To minimize wastage, fill trough feeders only two-thirds full. An inwardly rolled lip discourages billing out, as does raising the feeder so it's always the same height as the birds' backs.

By about their second week, the chicks will be ready for a bigger feeder. You'll know it's time to switch when chicks that climb through slots to eat get too fat to pop back out. The time to switch has passed when chicks are able to scratch strongly enough to overturn the feeder or they've grown too large to easily get their beaks into it. Whenever you change to a different feeder, leave the old one in place for a few days until you're sure all the chicks have learned to eat from the new one.

To minimize stress, chicks should drink soon after they hatch and eat within five hours.

As soon as our chicks are big enough, we like to switch them to a hanging tube feeder. A tube is the most versatile style of feeder because it can be used from about 1 week of age straight through to maturity. It can be hung by chains that simplify raising, and it holds lots of feed, so chicks are less likely to run out during the day. To determine how many chicks aged 6 weeks or under can be fed from one tube feeder, multiply the feeder's base diameter in inches (or centimeters) by 3.14 and divide by 2 (by 5 in metric).

Fill feeders in the morning, and let the chicks empty them fully before filling them again. Leaving feeders empty for long invites picking, but letting stale or dirty feed accumulate is unhealthful, so strike a happy balance. Clean and scrub feeders at least once a week.

Chick Problems

As chicks grow, cull out runts and deformed birds with crooked breasts or backs. Deformed chickens do not grow well, may not lay well, cannot be shown, and should never be kept for breeding. They can, however, be raised for family meat, although they're likely to grow less well than sound birds.

The typical death rate among chicks is 5 percent or less during the first 7 weeks, up to half of which are likely to occur during the first 2 weeks. Occasional deaths are therefore nothing to worry about. A pattern of increased deaths, however, is a sure sign of disease.

Disease is rarely a problem in newly hatched chicks unless they pick up an egg-transmitted illness during incubation, the incubator or brooder isn't kept properly clean, or the chicks are brooded too close to an adult flock, exposing them to a high concentration of microbes before they have a chance to develop natural immunity through *gradual* exposure. Because chicks are especially susceptible to adult illnesses such as Marek's disease and leukosis during their first 6 to 8 weeks of life, it's best not to brood them where adult birds have been raised in the past, or within 300 feet (90 m) of an adult flock. When you do chores, tend to your chicks before visiting older birds.

You can enhance the immunity of chicks with a probiotic consisting of the same variety of beneficial bacteria and yeasts that will eventually colonize their intestines and fend off harmful organisms through a process called "competitive exclusion." The intestines of artificially incubated chicks colonize more slowly than those of chicks hatched under a hen. To help things along, introduce beneficial micro-organisms early, letting the good bugs become well established so they can fight off bad bugs that might otherwise cause disease. Probiotic formulas come in two forms — for dissolving in water or sprinkling on feed — but both can be hard to find. A handy substitute is a little live-culture yogurt. Don't go overboard, though — giving chicks too much yogurt will cause diarrhea.

Stress due to chilling, overheating, dehydration, or starvation can drastically reduce the immunity of newly hatched chicks, making them susceptible to diseases they might otherwise resist. Routine activities such as vaccination or being moved also cause stress. To minimize the effects of stress, avoid exposing chicks to more than one stressor at a time. For example, try not to vaccinate, debeak, and move chicks to new housing all at once.

Omphalitis (also called "mushy chick disease" or "yolk sac infection") develops when bacteria invade a chick's unhealed navel. This illness occurs when high incubation humidity or low brooding temperature prevents the chick's navel from healing properly, leading to invasion by *E. coli* and other bacteria. The bacteria quickly multiply in the moist, nutrient-rich yolk sac area, causing chicks to die within a few days after hatching. Since there's no effective treatment, prevent this devastating disease by hatching only clean eggs, properly regulating your incubator and brooder, and practicing good incubator and brooder sanitation.

Pasting, also known as "sticky bottoms," is a common condition in newly hatched chicks. Although it can be caused by disease, it is more likely to be caused by chilling, overheating, or improper feeding. Soft droppings that stick to a chick's vent will harden and seal the vent shut, eventually causing death.

When pasting occurs, gently pick off the hardened droppings, taking care not to tear the chick's tender skin. If pasting continues, combine starter ration with an equal amount of cornmeal or crushed uncooked oatmeal, or feed scratch designed specifically for chicks, until the pasting stops. If you routinely feed chicks either chick scratch or scratch mixed with starter ration during their first two or three days of life, pasting is less apt to get started in the first place. Pasting in chicks older than one week is likely caused by an intestinal disease.

Coccidiosis, or "cocci" (coxy), is the most common intestinal disease among chicks. It results from protozoa that naturally colonize a chick's intestines but that multiply and get out of hand if the chick picks up too many protozoa by eating droppings in feed, water, or litter. Litter picking occurs when feed runs out or chicks are so crowded they either get bored (because they have too little room to move around) or they can't get enough time at the feeder.

Cocci is most likely to occur in chicks 3 to 6 weeks of age, with the worse cases appearing in birds 4 to 5 weeks old. It can be prevented with medications and by cleaning feeders and waterers often and keeping litter clean and dry. Coccidiosis is discussed more fully on page 276.

Starve-out occurs when chicks don't eat within two to three days of hatching. As a result, they become too weak to seek food. This problem may occur when:

◆ shipped chicks are in transit too long
◆ chicks can't find the feeders
◆ feeders are too high for chicks to reach
◆ excessive heat over feeders drives chicks away
◆ newly hatched chicks eat litter instead of feed

To keep chicks from eating litter, cover the litter with paper towels or some other rough material until they are eating well. Get them started by sprinkling a little feed on the paper towels or in paper plates. Place feed no more than 24 inches (60 cm) from the heat source but not directly under it.

Manure balls on toes invite toe picking and can cause crippling. Gently pick off the hardened droppings. If the toes bleed, rinse them with hydrogen peroxide and coat them with Neosporin antibiotic ointment. Be sure to correct the condition that caused the problem in the first place, most likely filthy brooder conditions due to crowding or improper litter management. Manure balls can also result from crooked toes that force a chick to walk on the sides of its feet.

Crooked toes can be caused by low incubation humidity or cold brooder floors, and for unknown reasons have been associated with the combination of infrared heat and wire mesh flooring. Other causes are vitamin E deficiency, injury, and heredity. You're most likely to see the condition in heavy breeds, especially in males. If you can't identify and correct the cause, don't keep crooked-toed chickens as breeders.

Toe picking and other forms of cannibalism have numerous causes. Among them: the brooder is too warm or too crowded, chicks run out of feed or haven't enough feeder space, the light is too bright, the ration is too low in protein. Solutions: decrease the brooder temperature more rapidly, increase the available space, increase the number of feeders, change to low-intensity or red lights, add 1 teaspoon of salt to each gallon of drinking water (1.3 ml per liter) or, better yet, switch to proper rations. Applying an antipick lotion, in my experience, is not a satisfactory solution.

Remove chicks that are visibly bloodied and raise them separately until their wounds completely heal. If picking continues, you may have no choice but to trim beaks. Permanent trimming — done by request by most hatcheries — disfigures birds (indeed, debeaked exhibition birds can't be shown) and makes preening difficult. To control cannibalism with temporary beak trimming, remove one-fifth of each chick's upper beak with nail clippers. The beaks will grow back in about 6 weeks.

Cockerels

At 3 to 8 weeks of age, depending on their breed, chicks start developing reddened combs and wattles. The cockerels' combs and wattles will become larger and more brightly colored than the pullets'. Unless they're Sebrights or Campines, which are hen feathered breeds, the males will soon develop pointed back and saddle feathers and long tail sickles, in contrast to the more rounded back and saddle feathers and shorter tails of hens.

At about the same time, peck-order fighting will get serious and sexual activity will start. If you haven't already done so, it's time to separate the cockerels from the pullets. Select the best cockerels for breeding and cull or butcher the rest.

Not long after their combs and wattles start to develop, cockerels will begin to crow. At first they'll sound pretty pathetic, more like a rusty hinge than a robust rooster. Some may end the sequence with a strangling sound. With practice, though, they soon get the hang of it, especially if a mature cock is nearby to set an example and provide competition.

Since cocks may crow at all hours of the night, the sound can become an issue with cranky neighbors. Generally, nighttime crowing is triggered by some disturbance, usually light or movement. Minimize crowing by keeping a rooster's quarters dark at night (including blocking windows through which passing cars may shine their lights) and by locating housing away from paths and driveways. If you can't keep light or sounds from entering your coop, block them out by leaving coop lights on all night and/or setting a radio on low. You can't eliminate crowing altogether, though, except through surgical "decrowing," an operation that's not always successful but is expensive *if* you can find a veterinarian who'll do it.

A cock inherits his style of crowing, so all cocks within one family sound somewhat similar. But each individual adds his own distinctive touch. If you have two or more roosters, you'll be able to recognize each one by the sound of his crow. I find it interesting that Americans use a five-syllable word, "cock-a-doodle-do," to describe the sound of a cock's crow. If you listen carefully, you'll more likely hear four syllables. In other languages, words used to describe crowing have four syllables: in French it's "coquerico," in Spanish "quiquiriquí," and in German "kikeriki."

CHAPTER 11

GENERAL MANAGEMENT

YOUR MANAGEMENT STYLE will depend on the type of chickens you raise, your purpose in keeping them, and the design of your facilities. Your management plan must include biosecurity measures and may also include litter management, integrating your flock with other livestock or with gardening, protecting your flock from weather extremes, and procedures such as claw trimming or spur removal.

Biosecurity

Biosecurity involves all the precautions you take to protect your flock from disease. For starters:

Provide a sound environment, as described in chapter 2.

Feed a balanced ration, as described in chapter 3.

Purchase only healthy stock. When possible, buy directly from a breeder rather than through an auction or other live-bird market where chickens from various sources are brought together.

Confine your flock. As picturesque as roaming chickens may be, getting hit by a car, eaten by a dog, or poisoned by lawn spray is bad for a bird's health.

Practice good sanitation. Unsanitary conditions are often caused by crowding, which forces chickens to perpetually live in manure. Remove manure piles and damp litter as they accumulate. Scrub waterers at least once a week. Clean feeders to prevent the accumulation of moldy feed.

Disinfect equipment used by other flocks. Any time you purchase used equipment or reuse equipment from a previous flock, scrub it well, disinfect it, and dry it in the sun before putting it to use.

Keep your chickens away from other chickens. Many diseases are spread by carriers — birds that transmit disease without themselves having symptoms.

Any time you introduce new chickens, you run the risk of introducing disease. If you acquire a new bird or bring one back from a show, isolate it for at least two weeks, until you're certain the bird is healthy.

Stay away from other flocks and keep other poultry people away from yours. This is one of the hardest biosecurity measures to observe, since those of us who enjoy chickens like to visit others with the same interest. To avoid spreading disease, cover your shoes (or visitors' shoes) with disposable plastic bags before going into the yard.

Don't mix birds of different ages. Salmonella, *E. coli*, and other bacteria may cycle back from older birds to younger ones, perhaps becoming stronger in the process; influenza can cycle forward from young birds to older ones.

Don't mix birds of different species. Germs that are relatively harmless in one species may have a devastating affect on another.

Keep wild birds away. Wild birds are especially likely to carry diseases — on their feet as they fly from flock to flock pilfering grain, in parasites on their bodies, or as illnesses they have in common with chickens. Screen windows and, if possible, place netting over your chicken run to keep out wild birds.

Control insects and rodents, both of which may bring in diseases. Keep your yard free of weeds and piled debris that attract vermin.

Medicate only when necessary. Giving your flock antibiotics or other drugs every time a bird looks droopy can cause stronger germs to develop that eventually won't respond to drugs at all.

Breed for resistance. Hatch chicks only from 100 percent healthy breeders. If you make it a practice to breed from birds that are at least 2 years old, you'll have plenty of time to weed out those that are more susceptible to disease.

Keep an accurate flock history. Should a problem arise, details that might come in handy include your birds' age, strain, and vaccinations; past diseases in your flock or on your place; contact with other flocks (by means of moved birds or human visitors); any suspicious symptoms you notice; and the circumstances surrounding deaths in the flock.

Burn or deeply bury dead birds and other animals. Chickens can get sick from picking at dead diseased birds or botulism-tainted animals (that perhaps died after wandering into the yard when hit by a car). Whether you burn or bury bodies will depend on local regulations.

Minimize stress. Chickens are always under stress in one form or another. To minimize stress, sing or talk whenever you approach your flock, always handle your birds gently, and avoid making more than one major change at a time. Treat your chickens with respect to help them remain calm, stress-free, and less likely to catch a disease. During times of unavoidable stress, including before and after a move or during unpleasant weather, boost their immunity with a vitamin supplement.

Stress Situations

Severity	Stressor
MINOR	Deficiencies in chicks due to inadequate breeder flock diet. Too much time between hatching and first feed or water. Cold damp floor. Debeaking. Handling. Low-grade infection. Consumption of spoiled feed or other toxic substances. Unusual noises or other disturbances. Extremely high egg production. Medication (severity depends on drug used).
IMPORTANT	Chilling or overheating during first weeks of life. Chilling or overheating of birds in transit. Sudden exposure to cold. Extreme weather and temperature variations. Extremely rapid growth. Unsanitary feeders, waterers, litter. Internal or external parasites. Insufficient ventilation or draftiness. Vaccination. Competitiveness between sexes and individuals.
SERIOUS	Overcrowding. Nutritional imbalance of feed. Dehydration due to inability to obtain drinking water.
SEVERE	Suffocation due to piling. Lengthy periods without feed or water. Starvation due to inadequate or poorly placed feeders. Contact among birds of various ages. Onset of any disease.

Adapted from: *Farm Flock Management Guide*, Floyd W. Hicks, The Pennsylvania State University.

Peck Order

Understanding peck order is important because the peck order governs a flock's social organization, thus reducing tension among the birds. In a flock containing both sexes, peck order involves a complex hierarchy on three different levels: among all the cocks, among all the hens, and between the cocks and the hens.

Some of the interesting things you'll learn by observing peck-order activities in your flock are that dominant cocks mate more often than lower-ranking cocks, but submissive hens mate more often than dominant hens because they crouch more readily. And, in a flock containing birds with various comb styles, single-comb birds are higher in rank than birds with other comb styles.

Chicks start sparring to establish the peck order at about 6 weeks of age. Once the peck order is established, fighting should be minimal, usually reserved to challenges to the top cock. The older he is, the more often he'll be challenged by a younger upstart. Studies at the University of Georgia suggest that you can reduce stress among your birds by helping your flock maintain a stable peck order.

- Give your chickens plenty of room to roam. Not only is exercise good for them, but the lowest ranking chickens need space to get away from higher ranking birds.
- Design housing with enough variety so timid birds can find a place to hide.
- Provide enough feeders and waterers for the number of chickens you keep. Otherwise, higher ranking birds will chase away lower ranking birds.
- In a mixed flock with more than one cock, position feeders and waterers so no bird has to travel more than 10 feet (300 cm) to eat or drink. Well-placed troughs allow each cock to set up his own territory and gather a group of hens around him; fighting is minimized if no bird has to pass through another's territory to reach feed and water.
- When you move chickens, do not combine birds from different groups. Doing so adds to the stress of moving and increases peck-order fighting.
- Avoid disturbing the peck order by introducing new chickens into the flock. Constantly introducing new birds causes stress that can lead to feather pulling, vent picking, and other forms of cannibalism.
- If you do introduce new birds, reduce bright lighting to make the unfamiliar birds less conspicuous.

Peck-order sparring starts at 6 weeks of age.

- If your chickens constantly fight or peck, look for management reasons such as poor nutrition, insufficient floor space, or inadequate ventilation.
- Never cull a bird just because it's lowest in rank — as long as you have at least two birds, one will always be lowest in rank.

Vaccination

Vaccination is a biosecurity measure you may or may not need. Newly hatched chicks have a certain amount of natural immunity, and they continue to acquire new immunities as they mature. Your flock may require help to develop additional immunities against diseases in its environment.

Ask your veterinarian or state Extension poultry specialist to help you work out a vaccination program based on diseases occurring in your area. Vaccinate your flock *only* against diseases your birds have a reasonable risk of getting, including past diseases birds have experienced on your place or new diseases that pose a serious threat in your area. Do *not* vaccinate against diseases that do not endanger your flock.

The most common diseases for which vaccines are available are: bronchitis, chronic respiratory disease, coryza, epidemic tremor, infectious bursal disease, laryngotracheitis, Marek's disease, Newcastle disease, pox, and viral arthritis. Some vaccines come in combinations. If you wish to administer two or more vaccines that don't normally come in combination, check with the supplier or a veterinarian to determine if the two are compatible. Some vaccines interfere with the effectiveness of others when used at the same time.

If you exhibit your chickens, you increase their risk of getting a disease. Although the chance is slim, there's always the possibility that some bird at a show will spread something, especially a respiratory illness such as laryngotracheitis (in the Northeast) or coryza (on the West Coast). Because some diseases are easily transmitted from bird to bird, and because recovered birds may continue to be carriers, exhibitors in some states are required to vaccinate against certain specified diseases.

Sample Vaccination Schedule

Vaccine	Age
Marek's disease	1 day
Newcastle disease, bronchitis	14 days
repeat in	4 weeks
and in	14 weeks
pox	8–10 weeks
epidemic tremor	14 weeks

Source: *Poultry Management*, H. Charles Goan, University of Tennessee.

Litter Management

Litter "management" began long before the idea caught the attention of university experimenters during the 1940s. Farmers, who have always been notorious for being behind in their work, tossed a layer of fresh bedding into the hen house whenever the place seemed a little messy. When they needed fertilizer, they spread the old litter on their fields and replaced it with fresh bedding. They didn't call it the "deep litter system" or the "built-up litter system" or any such fancy name.

During World War II, when livestock feedstuffs became scarce and expensive, the Ohio State Agricultural Experimental Station began seeking ways to reduce the need for animal protein in chicken feed. They discovered that decomposing litter is rich in vitamin B_{12}, a vitamin found only in animal protein, and a promoter of health, growth, and reproduction. It seems that chickens housed on naturally composting litter do not need a more expensive source of this vitamin in their rations. The "built-up litter system" was born.

Any good bedding (described on page 32) will work. Begin with a 4-inch (10 cm) layer. Whenever the surface gets matted, break it up and stir in a little fresh bedding until the layer is 10 inches (25 cm) deep by the start of winter. From then on, add fresh litter as needed to absorb the amount of manure your flock deposits. If your birds are not crowded, 12 to 15 inches (30–38 cm) of bedding should strike the right balance.

Rather than becoming filthy, as you might expect, properly managed built-up litter gradually ferments (composts) and after about 6 months develops sanitizing properties. Furthermore, the heat produced by fermentation keeps a flock warm during the cooler months, and flies are less of a problem in warmer months because accumulated dry manure attracts natural fly predators and parasites.

Managing built-up litter involves stirring the bedding as often as necessary to keep its surface from crusting over (a 4-prong cultivator is ideal for the purpose) and ensuring adequate ventilation so the litter will retain the right amount of moisture for good fermentation.

To test litter moisture, pick up a handful and squeeze. If the moisture level is just right, the litter will stick slightly to your hand but will break up when you let go. If it's too dry, it won't stick to your hand; if it's too wet, it will ball up and not easily break apart when you drop it.

If the bedding is either too damp or too dry, the coop environment becomes unpleasant and unhealthful for chickens and humans alike. Excessively dry litter not only fails to ferment properly (and therefore is not self-sanitizing) but also creates a dust problem. Dampen the litter with an occasional light sprinkling of water, followed by stirring.

Excessive moisture is more often a problem than dryness. Damp litter favors the growth of disease-causing molds and bacteria, and promotes the survival of viruses, parasitic worms, and protozoa (such as those causing coccidiosis). Damp litter also releases ammonia fumes that irritate avian (and human) eyes and respiratory tracts, opening the way to disease.

When litter is damp enough to emit ammonia, the first thing you'll notice is the odor. If your eyes burn and your nose runs, the ammonia level has become strong enough to increase your birds' susceptibility to respiratory disease. If the ammonia concentration gets so strong that birds' eyes become inflamed and watery, and the birds develop jerky head movements, ammonia blindness may soon follow.

To keep litter from getting too moist:

◆ Remove occasional damp spots that develop around doorways or waterers.

- Redesign entryways (as described on page 27) and fix dripping waterers to eliminate persistent damp patches.
- Repair roof leaks and correct drainage problems that cause indoor puddling.
- Insulate the ceiling, if necessary, to prevent winter condensation from dripping on litter.
- Add fresh litter more often or decrease the number of birds housed.
- Aerate the litter more often by loosening and turning it.
- Provide good ventilation to remove excess moisture from the air.

You can keep a flock on the same built-up litter for years, provided the bedding doesn't get damp and remains warm *and* no serious disease breaks out. On the other hand, if you need the nitrogen-rich bedding to fertilize your garden and/or summers in your area are quite warm, you may wish to clean the coop each spring and begin the summer with fresh, cool litter.

If frequent disease outbreaks in your area make the reuse of litter unwise or if you raise successive batches of birds for short production periods, use the "deep litter system": When you bring in a batch of new birds, spread the cleaned floor with 6 inches (15 cm) of fresh bedding. During the production period, stir the litter as necessary to prevent surface matting. At the end of each production period, remove the litter and thoroughly clean the coop.

Coop Cleanup

"Cleanliness is next to godliness, filthiness is next to death," said the hand-lettered sign tacked inside the coop of an old-timer I once visited. It was his daily reminder that safeguarding a flock's health is dependent on good sanitation. Sadly, sloppy sanitation is the most common cause of failure in raising chickens.

Good sanitation includes frequent cleaning of feeders and waterers, disinfection of reused housing or equipment, and regularly scheduled cleanup of house and yard. Unless a flock has experienced a health problem, a properly designed and maintained coop needs cleaning no more often than once a year. But even in the healthiest environment, disease-causing organisms build up over time. A thorough annual cleaning will remove 95 percent of the contamination.

To eliminate disease-causing organisms that flourish in the warm months of summer, schedule your major cleanup for spring. Choose a warm, dry, sunny day. Wear a dust mask so you won't breathe the fine dust you'll stir up. Lightly mist the walls and equipment to keep dust out of the air.

Remove all moveable feeders, waterers, perches, and nests. If you plan to remove the litter, now's the time to shovel it out. My favorite tool for moving loose litter is a snow shovel. If you'll be leaving the litter in place, shovel it back from the walls during cleanup. With a hoe, scrape caked manure from perches, walls, and nests. Use a shop vac or an old broom to remove dust and cobwebs from the walls, especially corners and cracks.

Mix 1 tablespoon (16 ml) of chlorine bleach per gallon (4 liters) of boiling water and use the solution to disinfect all troughs, perches, and nests, and to scrub the inside of the coop. Leave the doors and windows open to hasten drying. While you're waiting for the coop to dry, pick up any junk lying around the yard. Piles of scrap wood or discarded equipment keep out the sun's healing rays and attract insects, snakes, and rodents that cause birds harm.

Manure Management

Poultry manure is made up largely of undigested feed. For each 100 pounds (45 kg) of ration your flock eats, you can expect 45 pounds (20 kg) of droppings, dried weight. In addition to feed residue, fresh droppings contain intestinal bacteria, digestive juices, mineral byproducts from metabolic processes, and water. The white pasty stuff on top of a dropping is the avian equivalent of urine, consisting mostly of nitrogenous waste and water. Water makes up about 85 percent of the total weight of fresh chicken droppings, and its evaporation contributes to henhouse humidity, not to mention henhouse odor.

Pasturing chickens in portable housing is one way to avoid a manure problem — simply move the shelter often enough to keep manure from accumulating. In stationary housing, litter absorbs much of manure's moisture, minimizing both humidity and odor. Frequent manure removal also minimizes humidity and odors, but introduces new problems: it's labor intensive (who has time?); it requires a system for dealing with the manure; it can create a fly problem in warm weather.

Manure, left in one place over a period of time, attracts fly predators and parasites. When you remove the manure, you also remove the natural fly predators. With too few predators left to destroy them, fly eggs in the smallest clump of manure left behind will hatch into hundreds of flies. If you remove manure during the summer months, be prepared either to continue thoroughly cleaning your coop at least once a week until the fly season ends or to institute some other fly control measure, as discussed on page 280.

Minimizing Nitrogen Loss

The smell of ammonia coming from chicken manure — in either the henhouse or the compost pile — means nitrogen is evaporating. To reduce nitrogen losses, periodically apply ground rock phosphate or ground dolomitic limestone. Both substances combine with nitrogen to keep it from evaporating and to improve manure's fertilizer value. As a side benefit, they also help keep litter dry. For an average-size backyard coop, apply 1 pound (0.45 kg) per week stirred into litter or 2 pounds (0.9 kg) per week scattered over the droppings pit.

Manure as Fertilizer

Smart poultry keepers don't look at manure as a nuisance waste product but as a valuable commodity that can be used or sold for gardening. Aged manure and litter combined has average nitrogen (N), phosphate (P), and potash (K) values of 1.8, 1.4, and 0.8, respectively (as percentages of total weight), outranking most other barnyard droppings.

A good minimum yearly application is 45 pounds (approximately the amount produced by one hen each year) per 100 square feet of garden. A generous application is 90 pounds per 100 square feet. In practical terms, if you spread a 1-inch layer over 100 square feet of soil, you'll have applied roughly 55 pounds; a 1½-inch layer will give you just under 90 pounds. (In metric, a minimum application on a 10 sq m garden is 23 kg, a generous application is 45 kg, or about a 4.5 cm layer.)

Never feed growing plants fresh chicken manure — its high nitrogen content is hot enough to burn them. Furthermore, the excessive amount of nitrogen in fresh droppings encourages unbalanced plant growth such as weak plant stalks and forked carrot roots.

Farmers generally prefer to spread fresh manure over bare soil and turn it under at the end of the growing season in autumn, giving the manure time to age in the ground over winter. This practice, called "sheet composting," requires keeping chickens off the cropland for at least a year to prevent the spread of diseases and parasites.

Many gardeners prefer to compost manure for spring application to growing plants. Composting not only transforms nitrogen into a form that's readily usable without damaging plants but also destroys any bacteria, viruses, coccidia, or worm eggs that might be present. Built-up litter composts naturally, stabilizing or "fixing" both nitrogen and potash in the process.

Manure that's composted without being mixed with some carbon-containing substance (litter, weeds, grass clippings, etc.) will overheat and dry into a powdery white nutrient-poor ashy substance called "fire fang." If you're not an experienced composter, you can find all the information you need in any good book on organic gardening.

Fertilizer Value of Chicken Manure

The fertilizer value of chicken manure varies with its age and also with the nutritional intake of birds at various stages of growth.

	Nitrogen (N)	Phosphate (P)	Potash (K)
Chicken Manure, by degree of freshness:			
wet, sticky, caked	1.5%	1.0%	0.5%
moist, crumbly to sticky	2.0	2.0	1.0
crumbly but not dusty	3.0	2.8	1.5
dry, dusty	5.0	3.5	1.8
Fresh Manure, by age of birds:			
Baby chick	1.7	1.3	0.7
Growing chick	1.6	0.9	0.6
Hen	1.1	0.8	0.5
Comparison, barnyard averages (fresh):			
Cow	0.6	0.2	0.5
Horse	0.7	0.3	0.6
Steer	0.7	0.3	0.4
Duck	1.1	1.4	0.5
Goat	1.3	1.5	0.4
Turkey	2.0	1.4	0.6
Rabbit	2.4	1.4	0.6

Sources: USDA & The Pennsylvania State University.

Poultry and Diversified Farming

Excellent with Chickens	Reason
Cattle, Sheep, Meat Goats	Chickens obtain nutrients from undigested feed in livestock manure and keep down flies.
Horses	Chickens obtain nutrients from horse apples and make the horses less likely to spook.
Orchard	Chickens get exercise and fresh air while eliminating bugs and windfall fruit.
Pasture	Chickens benefit from sunlight, fresh air, and green feed while fertilizing the forage.

Good with Chickens	Reason
Garden	Chickens benefit from green feed and contribute fertilizer, but will damage plants.
Small Fruit	Chickens benefit from exercise and will keep down bugs, but will eat fruit.
Forest	Chickens benefit from fresh air and exercise, but attract woodlot predators.
Turkeys	Chickens gain resistance to Marek's disease from turkeys but also get blackhead from them.

Poor with Chickens	Reason
Dairy Goats	Chickens foul goat bedding and hay, and goats eat chicken rations.
Pigs	Chickens and pigs share susceptibility to avian tuberculosis.
Waterfowl	Damp conditions created by waterfowl are unhealthful for chickens.
Wetlands	Damp conditions are unhealthful for chickens, and water bugs transmit disease and parasites.

Chickens and Gardens

Chickens left to freely roam a garden will deposit N-P-K while destroying weeds and bugs, but they'll also eat new sprouts, scratch up seedlings, and peck holes in strawberries and tomatoes. Schemes for safely combining chickens with gardening are probably as old as chicken-keeping itself. Here are but a few of the many possibilities:

- Put the chicken house next to the garden, where you can easily toss plant refuse to the flock and collect nitrogen-rich manure for the compost pile.
- Surround the garden with a double-fenced chicken yard, or moat, creating a bug-free, weed-free zone that discourages entry by plant marauders including deer, rabbits, and groundhogs.
- Let chickens into your garden late in the day, giving them an hour or so to glean bugs and nip leaves but not enough time to do serious damage before they're ready to go to roost (keep them out while tomatoes are on the vine, though, as birds invariably make a beeline for ripe tomatoes).
- Keep a breed with heavy leg feathering, since they tend to scratch less than others and therefore do less damage to seedlings.
- Build a portable shelter to fit over raised beds so you can rotate the birds along with your veggies. Variations on this plan for enriching the soil to increase your harvest are discussed at length in Andy Lee's book *Chicken Tractor*, listed in the appendix.

Weather Considerations

A chicken doesn't need central heat and air to remain comfortable year-round. In cold weather, its body warms itself by producing approximately 35 BTUs per hour through physical activity and metabolic processes sustained by feed. In warm weather, its body cools itself by transporting internal heat to the external environment with the aid of its circulatory, respiratory, and excretory systems.

During long periods of extreme cold or heat, layers stop production and all chickens suffer stress. In general, chickens suffer less in cold weather than in hot weather, as long as their drinking water doesn't freeze and their housing is neither damp nor drafty. When temperatures reach 104°F (40°C) or above, chickens can't lose excess heat fast enough to maintain the proper body temperature, and deaths occur.

A chicken's body controls temperature by transferring heat between itself and its environment in four ways:

Radiation involves heat transfer between a chicken and nearby objects in its environment. If the environment is warmer than the bird's body, nearby objects warm the bird; if the environment is cooler, heat radiates from the bird to the environment. Examples of radiation control include: putting a reflective roof on the coop to prevent radiant heat gain from the sun in summer and insulating the roof to prevent radiant heat loss from birds in winter. At low temperatures, most of the heat a bird's body loses is through radiation and convection.

Convection is the transfer of heat between a bird and the surrounding air. Drafts and breezes cause warm air close to a bird's body to be replaced by cooler air. In winter, convection causes a bird to chill. At low temperatures, you can reduce convection by eliminating drafts that carry away warm air trapped by a bird's ruffled feathers.

In summer convection cools a bird, but only as long as the surrounding air is cooler than the bird's body temperature of 103°F (39.5°C). At air temperatures above 70°F (21°C), improve convection by opening doors and windows and, if necessary, by installing a fan. At 85°F to 90°F (29–32°C) a chicken exposes more of its body to moving air by holding out its wings.

Conduction is heat transfer between a chicken and objects its body contacts — floors, litter, nests, and so forth. As you might expect, warm objects warm the bird, cool objects cool it. Commonly, the parts of a bird's body having the most contact with external objects are its feet. Since its feet are quite small, their conductive influence is at best minimal. Another source of conduction is holes for dust baths: cool soil or fresh litter can be significant in keeping a bird cool in summer; warm composting litter provides warmth in winter.

Excretory heat transfer is a type of conduction that occurs when a bird drinks cool water, warms the water within its body, and eliminates the warm water in its droppings. At high temperatures, chickens increase the rate of heat loss by drinking more than usual, which causes their droppings to become loose and watery. Off-color droppings during the heat of summer may be a sign that birds aren't getting enough to drink. At high temperatures, excretion and evaporation account for most of a chicken's body heat loss.

Evaporation is the loss of latent body heat that occurs when the environmental temperature approaches a chicken's body temperature, and its body heat vaporizes liquid on the body's surface. Evaporation is an effective cooling method only when the air is low in relative humidity. Each 17°F (8°C) increase in air temperature doubles the air's capacity to carry moisture, up to a point —

air at any temperature can accept only so much moisture, and if the air is already saturated, it can't hold any more.

Respiratory heat transfer is a type of evaporation occurring when a bird inhales air that's cooler than the bird's body, and exhales moisture-laden warm air. The moist-air passages in the bird's extensive respiratory system — which include not only lungs but also air sacs among its organs and air spaces in some of its bones — help the bird lose internal body heat. In warm weather, a bird increases the rate of heat loss by panting. Since coops tend to be inherently high in humidity due to moisture produced by respiration and excretion, good ventilation and proper litter management are important temperature-controlling measures.

In summer, low humidity lets you take advantage of evaporation to cool birds by frequently hosing down the coop's outside walls and roof, and occasionally misting *adult* chickens, when the following conditions prevail:

♦ air temperature is above 95°F (35°C)
♦ air humidity is below 75 percent
♦ air circulation (convection) is good

In winter, humidity in the air chills chickens by drawing their body heat to the surface to vaporize. Furthermore, temperatures low enough to freeze moisture in the air can cause frostbite to combs, wattles, and toes. Frostbite is therefore more likely to occur in damp housing than in dry housing.

At any given time, a combination of all four forms of heat transfer determine whether a bird gains or loses body heat. A well-feathered bird, for example, will be comfortable at a temperature of 50°F (10°C), if the air is still and the sun is shining, while the same bird will be miserable at 68°F (20°C), if it's out in the wind and rain. Preventing either frostbite or heat stress therefore involves a combination of management measures.

Preventing Frostbite

To prevent frostbite:

♦ Reduce humidity by improving ventilation and removing damp patches of litter around doorways and waterers.
♦ Rake or pitchfork litter regularly — loose litter is dryer and has more insulating value than compact litter.
♦ Eliminate drafts by filling cracks and crevices in walls.
♦ Install perches in the least drafty part of the building.

◆ Use wide perches (at least 2 x 2 inches, 5 x 5 cm) that allow birds to cover their toes with breast feathers at night.

◆ Lower the ceiling to within 2 feet (60 cm) of perches to keep body heat close to the birds *or* install a heat lamp over the roost and plug it into a thermostat set to turn on the heat when the temperature drops below 35°F (2°C). Enclose the lamp in a sturdy wire guard so it can't be damaged in event of a collision with an airborne chicken. In a small coop, a few well-placed electric light bulbs may supply sufficient heat.

◆ Feed a small amount of scratch in the morning to kindle body warmth until birds are warmed by radiant heat from the sun.

◆ During cold days, stimulate appetites with a little mash moistened with warm skim milk or water.

◆ Increase interest in eating by feeding more often or by frequently stirring rations.

◆ Feed a little scratch at nightfall to increase body warmth during nighttime perching.

◆ Coat combs and wattles with petroleum jelly as insulation against frozen moisture in the air.

Reducing Heat Stress

Heat stress can be avoided by following these simple precautions:

◆ As water consumption goes up, increase the number of watering stations.

◆ Frequently fill drinkers with cool water.

◆ Keep water cool by placing waterers in the shade.

◆ Avoid medicating the water — if birds don't like the taste, they'll drink less.

◆ Since birds eat less in hot weather and rations go stale faster, ensure freshness by purchasing feed more often and in smaller amounts.

◆ Distribute feeders so birds don't have to travel far to eat.

◆ Encourage eating by feeding early in the morning and by turning on lights during cool morning and evening hours.

◆ Open windows and doors or install a ceiling fan to increase air movement.

◆ Eliminate crowded conditions.

◆ Do not confine birds to hot spaces such as trapnests or cages in direct sunlight or where ventilation is poor.

- ◆ Provide plenty of shade where birds can rest — if necessary, put up an awning or tarp.
- ◆ Do not disturb birds during the heat of the day.
- ◆ Hose down the coop roof and outside walls several times a day.
- ◆ Lightly mist adult birds (never chicks) when the temperature is high and humidity is low.

Weather Preferences by Breed

Cool

Brahma
Cochin
New Hampshire
Orpington
Plymouth Rock
Rhode Island Red
Wyandotte

Warm

Andalusian
Buttercup
Hamburg
Leghorn
Minorca
Naked Neck
Shamo
Spanish

Temperate

Cornish
Crevecoeur
Dorking
Houdan
Modern Game
Polish

Weather Preferences

Mature chickens can adapt to temperature extremes through gradual exposure. A slow steady shift in temperature therefore causes much less stress than a sudden change. When a bird becomes acclimated to warm temperatures, it pants less readily and is less likely to die at what might otherwise be a lethal temperature. By the same token, a bird that's grown used to warm weather is less tolerant of a sudden shift to cold weather.

Breed also plays a role in weather tolerance. Loosely feathered breeds like Orpingtons and heavily feathered breeds like the Asiatics (Brahma, Cochin, Langshan) and Americans (Plymouth Rock, New Hampshire, Rhode Island Red, etc.) suffer more in hot weather than lightly feathered breeds, and hens in lay suffer more than those not in production.

Cold weather affects lightly feathered breeds such as Hamburgs, Naked Necks, and the Mediterraneans (Buttercups, Leghorns, Minorcas, etc.), sparsely feathered breeds such as Shamo, and breeds with short close feathers such as Cornish and Modern Game. Not only are the latter two inadequately insulated against cold weather but both breeds do poorly in winter confinement.

Crested varieties (Polish, Houdan, Crevecoeur) are vulnerable to freezing if their copious head feathers get wet. Single-combed breeds suffer more in cold weather than rose-combed breeds; large-combed breeds like Dorkings suffer more than birds with smaller combs; cocks suffer more than hens (since they have larger combs and, unlike hens, don't sleep with their heads tucked under a wing).

The Chantecler, created in Canada as a dual-purpose breed with small comb and wattles, lays well in winter and can withstand cold weather but unfortunately is now nearly extinct.

Special Procedures

You may, at times, need to perform special procedures to solve specific management problems.

Dubbing and cropping, or the removal of a chicken's comb and wattles, has several purposes:

◆ keeps combs and wattles from freezing in cold weather (especially in cocks with large single combs)
◆ removes gangrenous tissue if frostbite does occur
◆ increases cold-climate egg production in hens with large single combs
◆ minimizes blood loss from wounds (especially among cocks that fight)
◆ satisfies requirements for showing Modern and Old English Game cocks

You can easily dub a day-old chick with small fingernail shears and cause little or no bleeding, but you won't get a show-quality job because the tiny comb offers little control. For show results, dub birds after their combs and wattles develop. A bird may be dubbed at any age, but younger birds (8 to 12 weeks old) bleed less than older birds.

In chickens of any age, the painful procedure is stressful and should be avoided in combination with other potential stressors. A vitamin K tablet, fed daily for 5 days preceding the operation, reduces bleeding by hastening clotting. A local anesthetic eases the pain. Dubbing and cropping are not operations you should perform without having first seen or helped an experienced person do it.

You'll need a pair of dubbing shears, which are basically 6-inch (15 cm) curved surgical scissors (available from any medical supply

Dubbing (removing the comb) and cropping (removing the wattles) may be necessary if a bird becomes injured or frost bitten.

store), or leather-cutting scissors (available from a harness or shoe-repair shop). Disinfect the scissors with alcohol between birds to avoid spreading bacteria. Have a helper hold each bird while you snip off its wattles one at a time, close to the neck. Then cut the comb ½ inch (12.5 mm) from the head. For a finished look, trim slightly closer to the head toward the front. To prevent infection, add tetracycline to drinking water at the rate of ¼ teaspoon per quart (1.3 ml per liter) for 10 days. The surgical wounds should heal within 30 days.

Beak trimming, or removing a portion of the upper and sometimes lower beak, is done to:

Beak trimming may be necessary to control cannibalism. Trim one-third of upper beak.

- ◆ reduce fighting among cocks
- ◆ stop egg eating among hens
- ◆ stop toe picking among chicks
- ◆ control cannibalism in intensely managed flocks

Although some hatcheries will debeak chicks on request, birds in a properly managed backyard flock should not need permanent debeaking. You may, however, occasionally need to trim beaks to nip a problem in the bud.

To trim a chick's beak, remove ⅕ of the upper portion with nail clippers. To trim a mature bird, use a knife to nick the upper beak ¼ inch (6.25 mm) from the end. With your thumb, hold the cut portion against the blade and roll the knife against the tip to tear the horny portion and expose the quick. In both cases, the beak should grow back in about 6 weeks.

Claw trimming is done to:

- ◆ keep cocks from injuring hens while breeding
- ◆ remove excessive nail growth of caged chickens
- ◆ groom birds for show

Trim toenails with a pair of canine toenail clippers or heavy shears, and file away sharp edges. If you inadvertently snip a nail too short, apply an astringent to stop the bleeding.

Spur removal is done to:

- ◆ prevent injury to a cock's human handlers
- ◆ prevent serious wounds to hens during breeding
- ◆ minimize injury in peck-order fights between cocks

- ◆ groom cocks for show
- ◆ remove excessive spur growth that may curl back into a cock's leg

For permanent removal, cut spurs off close to the leg when cockerels are 10 to 16 weeks old and their spurs are no more than ¼ inch (6 mm) long. Rub potassium hydroxide into the wound to prevent hemorrhaging and to destroy spur tissue so the spur can't grow back.

Trim the spurs of a mature cock in one of two ways: by snipping off the end and filing it smooth or by twisting off the spur cover. Beneath the old hard spur is a softer new one. If you choose to blunt the old spur by snipping off the end, take care you don't clip off too much at once or you might damage the new spur and cause it to bleed.

Exhibitors spruce up old show cocks by twisting off their spurs every couple of years.

*To remove a spur, first soften it with oil, then **twist** it back and forth until it comes loose; **never** bend the spur to break it off.*

First liberally apply oil to soften the juncture between the spur and the leg. Then *twist* the spur back and forth on its axis until it comes loose. *Do not bend the spur* to break it off or you'll likely damage the tender spur underneath, causing serious bleeding. After a few days, the new spur will harden. Meantime, confine the bird where he can't injure the freshly exposed soft spur tissue.

Controlling flight is used to confine chickens to their own yards for three main reasons:

- ◆ to protect chickens from predators
- ◆ to protect vegetable and flower beds from chickens
- ◆ to keep breeders from getting into the wrong pens

Breeds that are best known as flyers include Leghorn, Hamburg, Old English Game, and nearly any bantam breed. Heavier breeds may fly while they are young, but rarely after they fully mature. For breeds that don't fly high, a tall fence may be adequate. If the birds are kept in pens or runs, lightweight netting secured over the run will keep them in. Other flight-control methods involve altering a wing, either temporarily or permanently.

Two temporary methods — brailing and clipping — are used to control young birds that will eventually grow too heavy to fly. Brailing involves wrapping one wing with a soft cord so the wing can't be opened for flight. The cord, or brail, should be removed occasionally so it doesn't become tight enough to damage a growing bird and so the bird can exercise its muscles. Brailing might be appropriate for a show bird whose appearance would be marred by other flight-control methods.

Clipping, the most common method of controlling the flight of backyard chickens, involves using sharp shears to cut off the first ten flight feathers of one wing. Clipping causes a bird to lack the balance needed for flight but lasts only until new feathers grow during the next molt, which may be a few months in young birds or up to one year for older ones. A potential problem is that clipped feathers may not readily fall out during the molt, requiring your assistance.

Two permanent flight-control methods, pinioning and tenectomy, are more often used for game birds than for chickens but are equally effective for both. Pinioning, like clipping, controls flight by creating a lack of balance between a bird's two wings. It is a surgical procedure requiring sharp shears to remove the tip of one wing at the last joint. The operation causes the least amount of stress and bleeding if done when birds are no more than 1 day old.

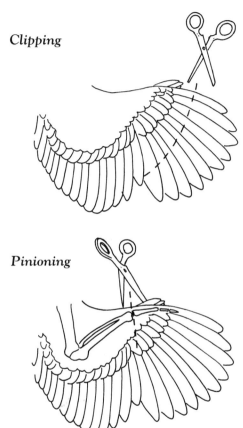

Clipping

Pinioning

Clipping the flight feathers is a temporary form of flight control. Pinioning, or removing the tip of the wing, is permanent.

Tenectomy calls for removal of a section of one wing's tendon — the tendon extending along the underside of the wing and running parallel to the major blood vessels — so the bird cannot control the wing for flight. It is the least often used method of flight control in chickens because it requires the knowledge of anatomy possessed by an avian veterinarian.

Catching Chickens

You'll need to catch your chickens to perform any of the above procedures, as well as to vaccinate them, move them, check for parasites, cull, and so forth. If you spend a lot of time taming your birds, catching one is as simple as bending down to pick it up. Otherwise, the easiest way to catch a chicken is to pick it off its perch at night.

In either case, reach beneath the bird to grasp it by its legs, then bring your other arm around to cradle it. Be sure to get hold of both legs or the chicken may churn around like an egg beater, possibly causing damage to itself or to you, and surely stirring up lots of dust. If you have to carry more than one bird at a time, turn them upside down and use their legs as handles. To avoid injuring birds, don't carry more than four in one hand.

If you have to catch a less-than-tame chicken during the day, you'll have an easier time with the aid of a fish net, catching hook, catching crate, and/or a helper. With a helper, slowly herd the bird into the corner of a building or non-electric wire mesh fence. When you and your helper get close enough, either quickly clamp down on both wings of the bird or reach beneath it and grab both legs. Take great care not to panic herded birds or cause them injury by letting them pile into each other or against equipment.

ALLAN DAMEROW

A hook lets you easily catch a chicken at any time of day.

Catching Hook

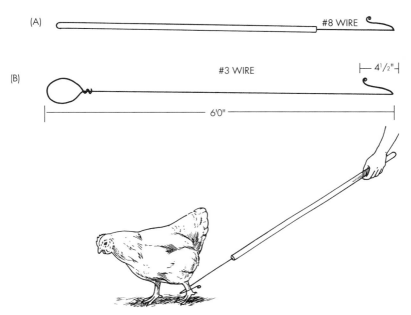

(A) *The standard catching hook consists of a 30-inch (75 cm) length of 8-gauge (4 mm) wire bent at one end into a hook and firmly attached at the other end to a wooden rake or broom handle. Less apt to frighten birds is (B) a 6-foot (180 cm) length of 3-gauge (6 mm) wire bent into one end and a circular handle bent into the other end. To catch a chicken, slip the hook over one leg.*

A catching crate helps you avoid exciting or injuring your birds. You can make one either from scratch or by modifying an existing cage or bird carrier. Despite its fancy-sounding name, a catching crate is nothing more than a lightweight but strong wooden or wire box with an opening at one end designed to fit against the door to the chicken's house or pen. The opening has a slide-down door you close after herding birds from the pen into the crate. A trap door at the top lets you handily reach in to get the bird you want. To round up chickens at pasture, use a pair of panels to guide birds into the open end.

If you have a fish net, you can catch a chicken by tossing the net over it, then retrieving the bird from beneath the net.

To use a catching hook, snare the bird by one leg. You can buy a hook from a poultry supply outlet or make one yourself. The standard hook consists of a 30-inch (75 cm) length of 8-gauge (4 mm) wire bent at one end into a hook and firmly attached at the other end to a wooden rake or broom handle. Less apt to frighten birds is a 6-foot (180 cm) length of 3-gauge (6 mm) wire with a hook bent into one end and a circular handle bent into the other end, as illustrated.

Any time you catch a chicken, take a moment to stroke its neck and wattles. By giving the bird a pleasurable experience, you'll have an easier time catching it in the future.

Transporting Chickens

A chicken can be safely transported in anything from a paper bag to a pet carrier, provided the container satisfies these criteria:

- ◆ The chicken can't get out.
- ◆ The container is not so big that the bird can hurt itself flying about, for example, if it's frightened.
- ◆ The container is not so small that the bird can't stand up and move around.
- ◆ The container has no sharp edges or other injurious protrusions.
- ◆ The chicken is protected from drafts, cold, and rain. (A stock rack or wire cage on an open pickup bed, for example, is not suitable for transporting chickens. On the other hand, a car trunk may not be suitable, either, if it accumulates carbon monoxide fumes.)
- ◆ The chicken has access to drinking water — if in transit, at least during occasional stops.

Determining Age

You can never be certain of a chicken's exact age (you can hardly, for example, check its teeth as you would a horse's), but you can always tell a young bird from an old one. First look the bird over. Cockerels and pullets tend to look like gangly teenagers compared to the more rounded, finished look of a cock or hen. Cockerels and pullets have smooth legs. Some pullets and all cockerels have little nubs where their spurs will grow. Older birds have rough scales on their legs. Cocks, and some hens, have long spurs. The longer the spur, the older the bird.

To confirm your findings, pick the bird up and examine it by feel. The breastbone is fairly flexible in a young bird, quite rigid in an older bird. The muscle is soft in a young bird, firm in an older bird. The skin is papery thin and somewhat translucent in a young bird, thick and tough in an older bird. A young bird will, in general, feel light compared to the solid, heavy feel of an older bird.

Predator Control

We humans aren't the only creatures fond of chicken meat, as confirmed by the profusion of poultry predators. Many of my neighbors profess to have bad luck when it comes to keeping chickens, but my flock has lost few birds over the years. One or two were carried away by hawks or eagles, and two were snatched one day by a family of foxes in the woods below our garden, where I'd foolishly let the chickens out to scratch.

In general, confined chickens are more susceptible to burrowing critters than pastured birds, while pastured flocks are more susceptible to flying predators. Many four-legged creatures don't like to expose themselves by crossing an open field, and they seem confused by or suspicious of housing that's moved every couple of days.

To deter predators, keep grass, weeds, and brush mowed around the henhouse. Place permanent housing on a solid foundation. Anchor portable housing with skirting that's tight and close to the ground; each time you move the shelter, double-check for dips where weasels can weasel in. A good fence, especially one that's electrified, will keep out most four-legged marauders, as will closing up the chickens at night. For a large flock, a guardian animal such as a dog or burro makes a good investment. If you opt for a canine, be sure he's reliable. A poultry breeder once remarked that her thriving repeat business was largely due to dogs — not just neighborhood dogs and marauding packs, but dogs owned by the same people who kept buying replacement chickens.

If you have a predator problem, you might call your local wildlife or animal control agency and see if they'll send out a trapper. Another option is to set a trap yourself. If you use a live trap with the intent of releasing the predator in some far off location, be aware that many animals are territorial and may eventually find their way back home. Others come in families, so catching one marauder won't necessarily solve your problem.

Electric fence baiting — a trick used to teach predators to respect a new fence — can be used with equal success to discourage a delinquent. If you find a dead chicken, hang it from a hot wire so the predator will get a shock when it comes back for another bite. If you don't use electric fencing, hang the chicken directly from an inexpensive fence controller. The bait should be nose height to the predator in a normal ambling stance, or about 6 inches (15 cm) above ground for a small creature to 3 feet (90 cm) for big dogs.

A predator-control option favored by many rural folks is to stand guard and shoot. If the marauder is your neighbor's dog, be sure to check local laws regarding your obligation to notify the neighbor about your intentions. If the predator is a wild animal that's protected by law, you're back to begging the wildlife agency for help. In our area, poultry people persistently complain that recently rein-troduced bald eagles are carrying off their birds; the wildlife people say "tough."

In an active chicken yard you're highly unlikely to find predator tracks before they're obliterated by chicken tracks, but each critter has a modus operandi that serves as something of a calling card to let you know what type of animal you're dealing with. Dogs kill for sport, then either chew up the chicken or just let it lie. I once had a carefree puppy that happily killed a dozen of my fryers and lined them up on the walkway in a neat row (end of puppy). Weasels and minks bite off a bird's head. Coyotes, foxes, bobcats, hawks, and eagles carry away the whole bird — it pays to count your chickens once in a while.

Rats generally chew up chicks and let them lie. Snakes eat chicks whole. We once found an otherwise harmless snake in our brooder after he had gulped down a chick and then curled up under the heat lamp to sleep off his fine meal. Opossums like tender growing birds and will sneak up to the roost while birds are sleeping to take a bite out of a breast or thigh.

Among egg eaters, a snake slithers into a nest to gorge on whole eggs, then either snoozes on the spot or else finds a tree to wrap around to break the shells, then sleeps off the meal nearby; the most likely snake to eat bird eggs is the rat snake. A jay or a crow pecks a hole in the shell, eats the contents, and leaves the empty shell in the nest (or sometimes carries it off and drops it at a dis-tance). A mink, opossum, raccoon, or "suck-egg" dog will leave telltale piles of shell in and around the nest. A rat takes the evidence with him by rolling intact eggs down his tunnel. A skunk, too, takes the whole egg, but leaves his odor behind. If you faintly smell skunk but find shell shards in or around the nest, you've likely been visited by an old boar raccoon.

Most predators work at night — some in the dead of night, others at dark or dawn. Exceptions are dogs (which make a meal any time they get the whim), coyotes (which hunt during the day only rarely), and foxes (which prefer to hunt early and late at night but will hunt during the day if game is scarce or they have young ones to feed). Among flying predators, great horned owls strike at night, while Cooper's hawks swoop down in daylight.

Rodents

Rodents are a particularly insidious type of predator. They're everywhere, they breed like rats, and they can't take a hint. They invade any time of year but get worse during fall and winter when they move indoors seeking food and shelter. Rats eat eggs and chicks, and both rats and mice eat copious quantities of feed and spread disease. Whether or not you find evidence, you can safely assume you have a rodent problem.

Discourage rodents by eliminating their hide-outs, including piles of unused equipment and other scrap. Store feed in containers with tight lids and avoid or sweep up spills. Aggressive measures include getting a cat, putting out traps, and — if you've got rats and you're experienced with a gun — shooting them. Don't bother with techie solutions like ultrasound "black boxes" and electromagnetic radiation — they're expensive and ineffective.

Poison bait stations work only if the rodents can find no other source of feed. Your choice of poison depends on how fast you want to kill the rodents and how safe the bait must be for the protection of children and pets. Poison comes in four types:

- Multiple-dose anticoagulants work slowly. A rodent has to nibble at the bait several times to be affected, so leave bait in place until no more is taken. Because rodents die slowly, they're likely to kick off in the wall or under the floor, creating quite a stench. But this type of poison is the safest to use. Trade names include Furmarin, Paracakes, Ramik, Rozol, and Warfarin.
- Single-dose anticoagulants work faster but are less safe for chickens and pets. They're still fairly safe, though, since an animal has to eat a massive amount to be poisoned. Leave the bait out for at least two days to make sure all rodents eat some. Trade names include Contrac, Havoc, Hawk, and Ropax.
- Cholecalciferol is a single-dose slow-acting poison that's relatively safe because any animal has to eat a large amount to be poisoned. Trade names include Mouse-B-Gone and Rampage.
- Bromethalin is a single-dose poison that's least safe to use because a small amount is toxic to all animals. Opt for one of the other three and leave this one alone.

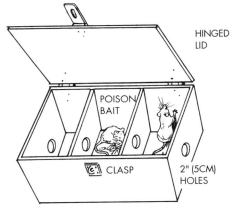

A wooden box a little larger than a cigar box makes an effective, pet-proof rodent baiting station.

CHAPTER 12

HEALTH MANAGEMENT

STARTING OUT with healthy chickens goes a long way toward keeping them that way. Maintaining a healthy flock involves recognizing and avoiding diseases, minimizing parasites, preventing poisoning, and managing the flock to discourage cannibalism.

Chicken Health

Chickens are susceptible to a large number of diseases, some mild, some devastating. Fortunately, most backyard flocks kept for family meat and eggs, or just for fun, rarely experience serious illnesses. Your flock may not be so lucky if other flocks are housed nearby, you habitually buy or trade chickens, you show your birds, or you run a small-scale commercial operation requiring a regular turn-over in your poultry population. If any of these situations pertain to you, you'd be wise to get a comprehensive health-care guide such as *The Chicken Health Handbook* (see Recommended Reading).

As a chicken ages, its state of health changes in two opposing ways:

◆ it develops resistance to some diseases in its environment
◆ it becomes susceptible to other diseases that require long-term development

From a health perspective, the older a chicken gets, the more valuable it becomes as a breeder, since by its longevity it demonstrates a certain hardiness that's likely to be inherited by its offspring. From a production standpoint, the older a chicken gets the less economical it becomes, with the profit curve dropping ever more sharply as long-term diseases take their toll. For the latter reason, many poultry owners won't keep each chicken more than a year or two.

Because commercial producers can't afford the economic loss resulting from a serious disease, and because mixing birds of different ages increases the likelihood of disease, commercial folks follow the "all in, all out" procedure — they keep a meat or egg flock for the most productive part of its life, then cull the whole flock and thoroughly clean the housing before bringing in another flock. So, although the natural life of a chicken is about 12 years — with individuals occasionally living as long as 25 years — rare is the chicken that survives beyond its productive life of about 2 years.

Well Chickens

The best way to nip a potential disease problem in the bud is to be constantly aware of your flock's state of health. Each time you visit your chickens, take a moment to look around. You'll readily spot problems in the making if you become fully familiar with these characteristics:

Appearance. Healthy chickens look perky and alert. They have full, waxy combs, shiny feathers, and bright eyes. Other signs of health are listed in the accompanying chart.

Activities. Healthy chickens are active. They peck, scratch, dust, preen, or meander almost constantly, except on hot days, when they rest in the shade.

Sounds. Well chickens "talk" and "sing" throughout the day. To detect atypical sounds, whistle softly whenever you approach your flock. Out of curiosity, your chickens will stop their activities to listen, giving you a chance to hear coughs, sneezes, and other unhealthy sounds.

Production. Each flock develops a characteristic laying pattern, based in part on the hens' breed and age and in part on your management style. Keep records to determine your flock's average rate of lay, as well as the seasonal variations you can come to expect. Familiarize yourself with the typical sizes and shapes of your hen's eggs. An inexplicable decline in production or in egg quality (thin or wrinkled shells, thin whites, and so forth) may be a sign of disease.

Consumption. The amount of feed and water a chicken consumes each day depends on its age, size, and production level, as well as on the weather. By paying attention to how much your chickens normally eat and drink, you'll readily notice changes brought on by stress or disease.

Weight. Healthy young chickens gain weight steadily. Healthy mature chickens hold their weight except for possible slight drops during the breeding season or resulting from the stress of exhibition. Any inexplicable loss of weight or failure to gain can be a danger signal.

Odors. A healthy flock has a characteristic odor. Become aware of that odor so you can detect subtle changes that may result from respiratory or intestinal disorders.

Droppings. Chickens expel two different kinds of droppings. Regular intestinal droppings are firm, grayish-brown, and capped with white urine salts. Approximately every tenth dropping comes from blind pouches in the intestine, called ceca, where cellulose is digested by fermentation. Cecal droppings tend to be somewhat foamy, smellier than regular droppings, and light brown or sometimes greenish in color. Any change in the normal odor or appearance of either kind of dropping is a pretty good indication of disease.

Signs of Health

Body Part	Health Signs
Comb and wattles	Bright, full, waxy
Eyes	Bright, shiny, alert
Nostrils	Clean, no breathing sounds
Head and tail	Held high
Breast	Full and plump
Abdomen	Firm but not hard
Posture	Erect, active, alert
Feathers	Smooth and clean
Vent	Clean, slightly moist
Droppings	Firm, gray-brown with white caps

From: *The Chicken Health Handbook* by Gail Damerow (Storey Publishing, 1994).

Treating Diseases

Avoiding diseases is both more effective and less frustrating than treating diseases, for these reasons:

◆ By the time you notice that a chicken is sick, chances are it's too far gone to be effectively treated.

◆ Continuing to house an unhealthy bird increases the risk that its disease will spread to others in your flock.

◆ Chickens that survive a disease rarely reach their highest potential as layers, breeders, or show birds.

◆ A chicken that fully recovers may become a carrier, continuing to spread the disease without showing symptoms.

◆ Some diseases are so serious that the only way to stop them is to dispose of the whole flock and start over.

Some diseases can be effectively treated if you catch them in the early stages *and* you can identify the disease so you'll know what treatment to use. A few illnesses have characteristic symptoms that, with a little experience, you can easily identify yourself. Many diseases have such similar symptoms that the only way you can get a positive identification, or diagnosis, is by taking a few dead or sick birds to a pathology laboratory for analysis. The path lab won't tell you how to treat the disease, though. For information on treatment, you'll have to take the path report to your regular vet or to your state Extension poultry veterinarian or Extension poultry specialist.

Develop a disease policy ahead of time so you'll be ready to act quickly if you ever have to deal with illness in your flock. The moment you suspect a chicken is sick, isolate it from the rest of the flock, then make a choice: either cull the bird and burn or deeply bury the body, or try to find out what the disease is and begin active treatment. If you wish to embark on a course of treatment, decide in advance how much time and money you are willing to spend nursing sick chickens.

Causes of Disease

Infectious (invasion by another organism)

bacteria
viruses
mold and fungi
parasites

Non-infectious (nonbiological in origin)

nutritional deficiencies
chemical poisoning
hereditary defects
unknown causes

Chicken Diseases

Infectious organisms of one sort or another are always present in the environment. Many of them don't cause disease unless a flock is stressed by such things as crowding, unsanitary conditions, or changes in feed. Microscopic organisms are spread through the air, soil, and water, as well as through contact with diseased chickens. They can be carried from flock to flock on the feet, fur, or feathers of other animals, especially rodents and wild birds, and on equipment, human clothing (particularly shoes), and vehicle tires.

The disease that's most likely to strike a specific flock is influenced by the flock's purpose. Exhibition birds are most likely to get an airborne respiratory disease such as laryngotracheitis. Broilers are most likely to experience diseases related to nutrition and rapid growth—ascites (fluid accumulation in the abdominal cavity), or leg weaknesses. Layers are most likely to experience a reproductive disorder such as egg-binding. A breeder flock is more likely than others to experience a disease requiring long-term development such as tuberculosis.

Each disease has unique symptoms by which it can be identified. Each disease also shares some symptoms with other diseases. General symptoms produced by nearly all infections include hunching, hanging of the head, dull or ruffled feathers, weight loss, and reduced egg production. Other symptoms can also be grouped according to the body system affected:

Enteric diseases affect the digestive system and are characterized by loose or bloody droppings, weakness, loss of appetite, increased thirst, dehydration, and weight loss in mature birds or slow growth in young birds. Enteric diseases include campylobacteriosis, canker, coccidiosis, colibacillosis (*E. coli* infection), necrotic enteritis, salmonellosis (typhoid, paratyphoid, pullorum), thrush, ulcerative enteritis, and internal parasites (worms).

Respiratory diseases invade a bird's breathing apparatus and are characterized by labored breathing, coughing, sneezing, sniffling, gasping, and runny eyes and nose. Respiratory diseases include coryza, cholera, chronic respiratory disease, bronchitis, gapeworm, influenza, laryngotracheitis, Newcastle disease, and wet pox.

Reportable Diseases

Some diseases pose such a serious threat to human health or to the poultry industry that they must be reported to state or federal authorities, who may allow you to treat your flock under supervision or may require disposal ("depopulation") of all the chickens on your premises.

Governing Body	Disease	Threat to
federal	chlamydiosis	human health
	exotic Newcastle disease	poultry industry
	influenza (lethal form)	poultry industry
	paratyphoid (*Salmonella enteritidis*)	human health
	spirochetosis	poultry industry
most states	pullorum	poultry industry
	typhoid	poultry industry
some states	chronic respiratory disease	poultry industry
	laryngotracheitis	poultry industry

Adapted from: *The Chicken Health Handbook* by Gail Damerow (Storey Publishing, 1994)

Nervous disorders affect the nervous system and are characterized by loss of coordination, trembling, twitching, staggering, circling, neck twisting, convulsions, and paralysis. Typical nervous disorders are botulism and Marek's disease.

Septicemia occurs when any infection reaches the bloodstream. Symptoms include weakness, listlessness, lack of appetite, chills, fever, dark or purplish head, prostration, and death. Many diseases have the ability to become septicemic — most notably cholera, colibacillosis, and streptococcosis (*Strep* infection).

Acute septicemia hits a bird so fast it literally drops in its tracks. Since most septicemic diseases cause reduced appetite and loss of weight before death, the classic indication of acute septicemia is sudden death of an apparently healthy bird that's in good flesh and has a full crop.

Not all causes of death are septicemic. Death can result from, among other things, degeneration of the intestine due to an enteric disease, blocking off of the airways due to a respiratory disease, inability to eat or breathe due to paralysis caused by a nervous disease, lack of adequate feed and/or water, or poisoning.

Finding a dead chicken is not necessarily a sign that a terrible disease is sweeping through your flock. Normal mortality among chickens is 5 percent per year. If you find several chickens dead within a short time, however, you most likely have good cause for concern.

Disease and Egg Laying

Reduced egg production is often the first general sign of disease, soon accompanied by depression, listlessness, loss of appetite, and weight loss. Diseases that affect laying, and their characteristic symptoms, include:

◆ Campylobacteriosis — shrunken combs, possibly diarrhea containing blood or mucous, sudden deaths; sometimes the only sign is a 35 percent drop in production
◆ Chronic respiratory disease — coughing, gurgling, swollen face, frothy eyes, sometimes yellowish droppings
◆ Infectious bronchitis — coughing; gasping; eggs with soft, thin, misshapen, rough, or ridged shells and watery whites; sharp drop in laying to nearly nothing
◆ Infectious coryza — watery eyes, swollen face, foul smelling nasal discharge, sometimes diarrhea
◆ Infectious laryngotracheitis — watery inflamed eyes, swollen sinuses, nasal discharge

- Lymphoid leukosis — no visible symptoms (internal tumors)
- Mild Newcastle disease — slight wheezing; eggs with soft, round, or deformed shells; temporary drop in production
- Paratyphoid — no symptoms other than a drop in production (signs of septicemia include purplish heads and sudden deaths)
- Salpingitis (oviduct infection caused by *E. coli*) — upright penguin-like posture, failure to lay, death
- Tuberculosis — gradual weight loss, prominent keel, pale combs, and persistent watery diarrhea in hens 2 years of age and up
- Visceral gout — depression, desire to hide, darkened head and shanks, white pasty diarrhea, drastic drop in production soon followed by death

Poisoning

Cases of chicken poisoning are few and far between. You shouldn't have a problem if you use common sense in handling or avoiding fungicide-treated seed (intended for planting), wood preservatives, rock salt, antifreeze, herbicides, rodenticides, and pesticides. Chickens can get poisoned not only from eating pesticides but from eating pesticide-killed bugs, as a friend learned when he lost his trio of pet bantams after they ate cockroaches he had sprayed outside his garage.

Chickens can be harmed by other foreign matter besides poisons. They peck at shards of glass, nails, staples, and other small, sharp objects that can cause internal injury. They are also attracted to string pulled from feed sacks, which can wrap around a tongue and cause choking. All these situations can be easily avoided with a little forethought.

Potential Poultry Poisons

Poison	Source
carbon monoxide	carrying chickens in car trunk
copper sulfate	antifungal treatment
ethylene glycol	spilled antifreeze
lead	paint or orchard sprays
mercury	disinfectants and fungicides
rock salt	de-icing sidewalks

Parasites

A parasite is, technically speaking, any living thing that invades the body of another living thing and takes sustenance without providing benefit. To a specialist, all infections are caused by parasites of one sort or another. To the nonspecialist, a parasite is an animal form (such as a worm or a mite) that lives on or within another animal form (such as a chicken). This common definition excludes infections caused by parasitic plant forms — bacteria, viruses, and fungi.

Although animal-form parasites are common among chickens, not all pose a serious threat. A heavy infestation of even the most benign parasite, however, causes stress that lowers a chicken's susceptibility to other infections. Parasites are spread by wild birds, rodents, and chickens. They can be brought into a flock on used feeders, waterers, nests, and other equipment that hasn't been thoroughly cleaned before being put to use.

Which parasites are most likely to infect your flock depends in part on your management style and in part on the area of the country in which you live. Your local vet or state Extension poultry specialist can tell you which parasites to watch out for.

Internal Parasites

Internal parasites live inside a chicken's body, usually in some part of its digestive tract. The main internal parasites that invade chickens are worms and protozoa.

Worms

Two categories of worm infect chickens: roundworms and flatworms.

Roundworms (nematodes) are the most significant parasitic worm in terms of the number of species involved and the damage they do. The most likely nematodes to infect a chicken are:

◆ *cecal worms* — short worms that invade the ceca (two blind pouches attached to the intestine), causing either no symptoms or weight loss and weakness. Although these are the most common of all nematodes, they rarely cause serious problems.

◆ *large roundworm* (ascarids) — long, yellowish-white worms that invade the intestine, causing pale heads, droopiness, weight loss, diarrhea, and death. Chickens usually become resistant to ascarids by 3 months of age.

◆ *capillary worms* — hairlike worms that invade the crop and upper intestine, causing droopiness, weight loss, diarrhea, and sometimes death. When chickens sit around with their heads drawn in, capillary worms are the likely culprit.

◆ *gapeworm* — red fork-shaped worms that invade the windpipe, causing gasping, coughing, and head shaking (in an attempt to dislodge the worm). These parasites are quite serious in young birds, potentially causing death through strangulation.

Flatworms fall into two categories:

◆ *tapeworms* (cestodes) — long white ribbonlike segmented worms that invade the intestine, causing weakness, slow growth or weight loss, and sometimes death. Although tapeworms infect some 50 percent of all flocks, they are not usually serious.

◆ *flukes* (trematodes) — broad leaf-shaped worms that attach themselves either inside the body or beneath the skin. Flukes are a problem primarily in swampy areas and where sanitation is abysmal.

Deworming

Some poultry keepers periodically deworm their flocks without knowing whether or not their chickens have worms, a practice that wastes money and may actually be harmful. Indiscriminate deworming is a bad idea because:

◆ it can cause drug-resistant parasite populations to develop
◆ it does not allow chickens to acquire resistance to parasites

A bird that has the opportunity to acquire resistance through gradual exposure to the worms in its environment gets an unhealthy worm load only if it is seriously stressed by such things as crowding, unsanitary conditions, or the presence of some other disease.

Rather than deworming indiscriminately, a more sensible approach is to have fecal samples examined periodically for signs of worms. Any veterinarian will do a fecal test for a few dollars. When you take in a sample of fresh droppings (feces), the vet will tell you if your chickens have a serious worm load and, if so, what kind of medication will rid your birds of the species involved.

By taking fecal samples to your vet on a regular basis (perhaps every three months for a year), you can find out whether or not your chickens have a serious parasite problem and whether the problem varies in severity with the season. Then you can develop a deworming schedule based on your flock's need, rather than on the biased advice of dealers who make a living selling poultry drugs.

Intermediate Hosts

To effectively control parasitic worms, you have to know something about their life cycles. Some have a direct life cycle; others have an indirect life cycle.

In a *direct life cycle,* a female worm inside a chicken's body sheds eggs that are expelled in the chicken's droppings. The infective parasite egg may then be eaten by the same chicken or by a different chicken. Assuming the parasite is a "guest" of the chicken it invades, the chicken becomes its "host." A direct-cycle parasite goes directly from one host chicken to another.

In an *indirect life cycle,* worm eggs expelled in the chicken's droppings are eaten by some other creature such as an ant, a grasshopper, or an earthworm. A chicken cannot become infected (or reinfected) by eating a parasite egg, but rather becomes infected (or reinfected) by eating a creature containing a parasite egg. Because the parasite goes from a chicken to some other host and back to a chicken, the other creature is called an "intermediate host." Parasites requiring an intermediate host are said to have an indirect life cycle because they cannot infect one chicken directly after leaving another. Most roundworms and all tapeworms are indirect-cycle parasites.

The type of intermediate host a chicken is most likely to be exposed to, and therefore the type of parasite most likely to infect the bird, is partly determined by the bird's style of housing. Free-ranged chickens are most likely to be infected by parasites whose life cycles involve ants, earthworms, slugs, or snails. Litter-raised flocks are likely to be infected by parasites involving beetles, cockroaches, or earthworms (or by direct-cycle parasites). Chickens housed in cages are likely to be infected by parasites whose life cycles involve flying insects.

Most intermediate hosts proliferate during warm weather. Frequent deworming may be needed in a warm, humid climate where intermediate hosts thrive year-round. In northern areas, where intermediate hosts thrive in summer but become dormant in winter, parasite loads may only be large enough to require deworming once a year, in autumn.

Parasitic Worms and Their Hosts

Worm	Intermediate Host
ascarid	none (direct cycle)
cecal	none or beetle, earwig, grasshopper
capillary	none or earthworm
gapeworm	none or earthworm, slug, snail
tapeworm	ant, beetle, earthworm, slug, snail, termite
fluke	dragonfly, mayfly

If you prefer to avoid drug use, sooner or later someone will tell you that the best way to keep chickens free of worms is to feed them diatomaceous earth — diatom fossils ground into an abrasive powder that shreds delicate bodies. I have never seen evidence that diatomaceous earth is an effective dewormer, and common sense tells me that it couldn't be. When diatomaceous earth gets wet (as it would inside a chicken's digestive tract), it softens and loses its cutting edge.

The best way to control worms is to provide good sanitation and control intermediate hosts.

Coccidiosis

Protozoa are the simplest and smallest members of the animal kingdom. Some are harmless, others cause serious illness. The protozoa most likely to infect chickens are coccidia, the most common cause of death in young birds. Although many different animals can be infected by coccidia, the species that infect chickens do not affect other kinds of livestock. The opposite is also true — other animals cannot get coccidiosis from a chicken.

Coccidia have direct life cycles. For each egg that hatches in a chicken's intestines, millions are later released in the bird's droppings. All ground-fed birds are exposed to coccidia throughout their lives. A properly maintained free-ranged flock is less likely to become infected than birds living in crowded conditions, housed on damp litter, or allowed to drink feces-fouled water.

Gradual exposure to coccidia allows a chicken to develop immunity. In young birds that are not yet immune, illness or death occurs when poor sanitation exposes them to too many coccidia too rapidly. Chicks raised on wire and later moved to litter have had little exposure, and therefore have no immunity and can become seriously infected. Even mature birds can become infected in hot, humid weather when coccidia proliferate more rapidly.

Nine different forms of cocci are caused by nine separate species of protozoa, each invading a different part of the intestine. A chicken can be infected by more than one species at a time. Birds become immune in two ways: through gradual exposure or by surviving the illness. But they become immune only to the species occurring in their environment. Healthy chickens brought together from different sources may not all be immune to the same forms, and may therefore transmit the disease to one another with devastating results.

In young chickens, the main symptoms of coccidiosis are slow growth and loose, watery, or off-color droppings. If blood appears in the droppings, the illness is serious — birds may survive but are unlikely to thrive. Sometimes the disease develops slowly, sometimes bloody diarrhea and death come on fast. In mature birds, the chief sign is a decrease in laying. Infected older birds that appear healthy shed billions of eggs that readily infect younger birds.

If you suspect coccidiosis, take a sample of fresh droppings to your veterinarian and ask for a fecal test to find out what kind of coccidia are involved and which medication to use. Not all anti-coccidials work against all types of coccidia, and using the wrong medication can do more harm than good.

Chicks reared early in the year while the weather is still cool are in the best position to develop immunity. To prevent an outbreak in late-hatched chicks or chicks raised where the climate is warm year-round, feed medicated starter ration containing a coccidiostat. A coccidiostat designed to either prevent or control coccidiosis can also be added to drinking water, but use it with caution during warm weather — birds drink more when it's hot and can obtain a toxic dose.

External Parasites

External parasites that live on or attack the outside of a chicken's body include mites, lice, and a host of fleas, flies, and other minor pests. An infestation causes irritation at best; a serious infestation can cause death.

Check your chickens for external parasites at least once a month. Since they spread rapidly, you needn't check more than a few birds. To catch parasites that attack only at night, examine birds after dark, using a flashlight.

Most external parasites can be prevented with good management, starting with proper sanitation. Once an infestation occurs, the only effective recourse may be to use an insecticide. The list of insecticides approved for poultry is short and changes often, so check with your state Extension poultry specialist or veterinarian for the latest information. Some approved insecticides are so potent you need a permit to use them. Never use a nonapproved product on chickens raised for meat or eggs. Even an approved insecticide is toxic and must be handled with care, so read the label and follow all precautions.

Mites

Several species of mite live on the skin, feathers, or blood of a chicken. Mites cause irritation, feather damage, increased appetite, low egg production, reduced fertility, retarded growth, and sometimes death.

Red mites are active in warm climates during summer, and are especially attracted to broody hens. They invade chickens primarily at night, when they appear as tiny specks crawling on roosts or on a bird's body. Control them by thoroughly cleaning the coop and dusting every crack and crevice with an approved insecticide.

Northern fowl mites are active in cool climates during winter, causing scabby skin and darkened feathers around the vent. You'll know you have an

Parasite Zappers

Dust baths in dry soil or fine road dust, and the preening that invariably follows, help a chicken rid its feathers of parasites. To further control parasites, old-time poultry keepers laced their dust bins with wood ashes, diatomaceous earth, or lime-and-sulfur garden powder. But chickens are highly susceptible to respiratory problems, and breathing these foreign materials can make matters worse. On the other hand, a parasite-infested bird will likely benefit more from being rid of parasites than it would be harmed by inhaling exotic dust.

Nicotine sulfate (a gardening pesticide sold under the trade name Black Leaf-40) is another old-time control method that works well but is so toxic it's banned in some states. Some poultry keepers paint it on roosts; others apply it to vent feathers *(not* to the skin, which can kill a bird). Warmth from the bird's body causes toxic fumes to evaporate, permeate the feathers, and kill body parasites.

Systemic inhibitors permeate a chicken's entire system, making its skin and blood unappealing to external parasites. The coccidiostat sulfaquinoxaline acts as a systemic inhibitor. So does ivermectin, an over-the-counter cattle dewormer that is not approved for poultry and should not be given to meat or egg birds. Sold under the trade name Ivomec, ivermectin is given by mouth, ¼ cc per chicken, 5 to 7 drops per bantam. Sulfaquinoxaline and ivermectin are both toxic to birds in excessive doses.

Petroleum oil — applied to roosts, nests, and cracks in walls or floors — rids a coop of parasites that spend part of their time off a bird's body. Oil is effective but messy, and can cause a fire hazard in a wooden building.

Pet shampoo and flea dip make excellent parasite control options, especially if you wash your chickens for exhibition.

infestation if you see tiny specks crawling on eggs in nests or on birds during the day. These critters increase rapidly, so act fast by dusting birds and nests with an approved insecticide. Wise old poultry keepers discourage mites by lining nests with tobacco leaves or by using cedar chips as nesting material.

Scaly leg mites burrow under the scales of a chicken's legs, making the scales stick out and causing the miserable bird to walk stiff legged. Since leg mites travel slowly from one bird to another, they can be easily controlled by brushing perches and birds' legs once a month with a mixture of one part kerosene to two parts linseed oil *(not* motor oil). Older birds are more likely to be infected than younger birds and are difficult to treat because the mites burrow deeply.

The traditional treatment is to smother the mites by coating legs of infected birds with petroleum jelly or a kerosene/oil mixture. The most effective treatment is ivermectin, used as described in the accompanying box "Parasite Zappers."

Chiggers are prevalent in eastern, southern, and southwestern states. They pierce the skin of a chicken's neck, breast, and wings, causing tiny itchy scabs that irritate for weeks. A serious infestation can make a bird so miserable it stops eating and eventually dies. Since chiggers are the larvae of plant-eating mites, the best way to control them is to mow weeds around the coop and, if the situation is extreme, dust or spray the area with sulfur.

Ticks are basically big mites. The ones most likely to bite chickens are the fowl tick and the lone star tick.

The fowl tick is found in the Southwest and along the Gulf coast. Its color is tan or reddish brown until it feeds on blood, which makes it turn bluish. Larvae attach themselves and feed for about a week; adults feed for only about 15 to 30 minutes at night. Chickens roosting in an infested coop get restless at nightfall because they expect to get bitten.

Controlling fowl ticks entails providing facilities in which ticks can't easily crawl onto birds. A metal coop, suspended roosts, and wire cages all inhibit ticks but won't stop them. In a tick-infested area, remove trees close to the coop (since ticks lay their eggs in tree bark) and spray housing once a month with an approved insecticide. If an infestation occurs, you have two choices: treat each bird and spray the housing every week for eight weeks *or* get rid of the infested chickens, burn the house down, and start over.

Lone star ticks live in woodland and brushy areas of the Southeast, where they primarily attack free-ranging chickens. The adult female is pear-shaped and chestnut-brown with a pale spot on her back; larvae are straw-colored and crawl around in bunches; nymphs look like tiny adults, only are lighter in color. These ticks bite humans as well as chickens, and transmit diseases. Control them by keeping your chickens out of brushy areas and by cutting down brush around the coop.

Lice

Most lice don't harm chickens because they eat feather parts, dead skin scales, and other debris on the skin. Two kinds of lice — body lice and head lice — chew on a chicken, causing the bird to break off or pull out its feathers in an attempt to stop the irritation. Lice also cause a drop in laying and fertility.

You can easily see the tan or transparent-looking pests scurrying around on a chicken's skin, the scabby dirty areas of skin they create around the vent and tail, and louse eggs (called "nits") clumped in masses where feather shafts come

Signs of Common External Parasites

Parasite	Symptoms
Red mites	red specks crawling on skin at night, weight loss, death
Northern fowl mites	red or black specks around vent or on eggs in nests, weight loss, drop in laying
Scaly leg mites	enlarged shanks and toes with raised, crusty scales
Chiggers	tiny red scabs on skin, loss of appetite, weight loss, death
Ticks	ticks on body, infected bites, ruffled feathers, loss of appetite, drop in laying, weakness, death
Lice	pale insects scurrying on skin, eggs clumped at base of feathers, dirty looking vent and tail area, weight loss, reduced laying and fertility
Sticktight fleas	clusters of reddish-brown specks attached to head

out of the skin. Lice spread through contact with an infested bird or its feathers, and they live their entire lives on a bird's body. If you spot signs on one bird, treat them all. Since insecticides won't kill louse eggs, repeat the treatment two more times in seven-day intervals to kill lice that hatch between treatments.

Sticktight Fleas

Sticktight fleas are common in sandy areas across the southern United States and northward into New York. They attach in clusters to the skin of a chicken's head. Although they may be easily controlled by applying a flea salve to an infected bird's head, the fleas are difficult to remove from housing. The moment you spot sticktight fleas, remove all litter and *heavily* dust the floor with a poultry-approved insecticidal powder. Repeat the dusting two or three times at 10- to 14-day intervals.

Flies

Flies that bother chickens fall into two categories: biting flies and filth flies. *Biting flies* include black flies and biting gnats (midges, no-see-ums,

HEALTH MANAGEMENT **281**

punkies, sandflies) found primarily around bodies of water. Their bites irritate chickens and can spread disease. Control biting flies by keeping chickens away from streams and stagnant water.

Filth flies, including the common house fly, don't bite, but they transmit tapeworm (when eaten by a chicken) and spread diseases on their feet. Since flies breed in damp litter and manure, control them by keeping litter dry — fix leaky waterers and roofs, regrade to prevent run-off seepage around the foundation, and improve ventilation. If flies get out of control, set out fly traps or a good fly paper such as Sticky Roll, or introduce natural fly predators. If you let litter and manure accumulate over the summer, it will develop its own population of fly predators. Avoid using an insecticide or you'll likely end up with a resistant fly population.

Cannibalism

Cannibalism is the nasty habit chickens have of picking at one another. Leghorns and other light, high-strung breeds are more likely to engage in cannibalism than the heavier, more sedate American and Asiatic breeds. The habit usually starts with one bird and spreads to others. Identifying and removing the culprit often stops the problem before it really gets rolling.

Early forms of cannibalism among chicks include toe picking (a chick's toes look remarkably like little worms) and feather picking (chickens love tasty red treats like strawberries, ripe tomatoes, and newly emerging blood-filled feather quills). Among chickens of all ages, bleeding injuries can lead to cannibalism. So can overcrowding, boredom, lack of exercise, too few feeding or watering stations, feed or water troughs too close together, high-calorie low-fiber rations, bright lights, excessive heat without proper ventilation, nutritional imbalance (too little salt or protein), and external parasites.

Forms of Cannibalism

Form	Likely Group
toe picking	chicks
tail pulling	growing birds
feather picking	growing birds
vent picking	pullets
egg eating	hens
head picking	cocks, any birds in adjoining cages

Cannibalism is a management problem. Prevent it among chicks by avoiding crowding, reducing brooding temperature, and increasing ventilation as the chicks grow. Alleviate boredom by letting chickens run outside where they have plenty to peck besides each other.

Egg eating is a form of cannibalism that usually starts when eggs get broken in the nest (see discussion of broken shells on page 123). Once chickens find out how good eggs taste, they break them on purpose to eat them. The only way to stop egg eating and keep it from spreading is to remove the culprit early. Identify the instigator either by checking for egg yolk smeared on beaks or by catching the eater in the act.

CHAPTER 13

SHOWING

HOW WELL YOUR BIRD places in an exhibition depends on its condition, its disposition, how closely it conforms to the *Standard* description for its breed and variety, how it compares with other birds in its class at the show, and the adeptness of the judge who reviews its class. How well you do as an exhibitor — and how much enjoyment you get out of it — depends on your reasons for showing and how well prepared you are for each event.

Why Show?

People who show poultry are called "fanciers" because they fancy, or are fond of, their chickens. They enjoy showing birds for five main reasons:

- *To get feedback from judges and other exhibitors.* Serious breeders enjoy the challenge of constantly trying to improve their breeding stock and their showing skills. They like to compare the results of their breeding and conditioning programs with other entries, and they enjoy the opportunity to talk shop with other experienced breeders at a show.
- *To win awards.* Winning can be the result of experience and skill at breeding and conditioning birds, or it can be the result of acquiring the best birds money can buy. Those with experience and skill tend to win consistently. Those who habitually purchase the birds they show tend to be inconsistent winners and poor losers.
- *To sell birds.* Not all exhibitors are interested in selling birds, and not all shows allow selling on the premises. Where selling is allowed, serious breeders may sell an occasional bird to help someone get started in their favorite hobby or to help defray the cost of showing.

Some exhibitors enter shows solely to advertise and sell large quantities of birds, which tend to go for lower prices than birds sold by serious breeders because they are generally of inferior quality.

◆ *To satisfy some noble cause.* A minority of exhibitors enter shows to serve some higher purpose, such as attempting to arouse interest in a breed or variety they fear may be nearing extinction or to educate the general public about the benefits and joys of keeping chickens.

◆ *For the camaraderie.* A few exhibitors simply enjoy chickens and like to visit with others who also enjoy chickens. Many (but not all) of those who show chickens solely for sociable reasons are novices.

Selecting Show Birds

Many novices start out trying to show too many different breeds or varieties, and as a result don't do well with any of them. These newcomers have the mistaken idea that the more birds you enter, the better chance you have of winning. But showing chickens isn't so much a game of chance as a game of skill. The first "secret" to success is to specialize.

Those who win consistently are exhibitors who are also breeders with an indepth knowledge of the genetics of their chosen breed and variety. For them, showing starts in the breeding pen: making carefully thought-out matings, watching young birds grow, culling those with deformities or with incorrect type or color.

Since type defines breed, type comes before color in selecting a show bird. Type includes not only a bird's overall weight and size but also the shape of its head, the slope of its back, the carriage of its tail, the breadth of its stance, the quality of its feathering, and a myriad of other details that characterize each breed, as described in the *Standard* — the exhibitors' bible.

The *Standard* also describes the proper color that defines each variety. Color includes not just the appearance of visible plumage but the color of the underfluff and skin underlying the plumage, as well as the comb, shanks, ear lobe, eye, and skin surrounding the eye.

Since preparing a bird for show takes time, it pays to condition only your best birds.

Conditioning

Conditioning is the process of bringing a show bird to the peak of cleanliness and good health. A properly conditioned bird has that undefinable quality known as "bloom." No matter how much time you spend with a bird, you can't keep it in a constant state of bloom. All birds are unsuitable for showing, for example, while going through their annual molt. Furthermore, constant showing causes a bird stress, and stress is not conducive to bloom (or to good fertility in the breeding pen). If you wish to attend a large number of shows, rotate the birds you condition.

In mapping out your conditioning program, obtain copies of the premium lists for the shows you wish to attend. The premium list outlines the classes and varieties that will be accepted for each particular show. Not all birds are eligible for all shows. Some shows allow only large breeds or only bantams; others allow only certain breeds or varieties (due perhaps to space limitations or to specific interests of the show's sponsors). If you have any doubts, call the show superintendent or secretary.

Other information you'll find in the premium list includes the show's date(s), the deadline for sending in your entry form, the cost of entering, health requirements (vaccinations, blood tests, health certificates, and so forth), and the prizes or premiums being offered. The value of the prizes often determines the amount of competition you can expect in your class. Competition will likely be stiff if your class involves a sanctioned meet — prizes offered by a specialty club for its members. To be eligible for such prizes, you must join the group. Joining the specialty club for your breed is a good idea, in any case. By reading the group's newsletter and talking to other members, you will learn about problems that are peculiar to your breed.

Once you have determined that your birds are eligible for the show you wish to enter, examine each bird again to make sure it won't be disqualified for any of the numerous reasons stated in the *Standard*. Check once more for hereditary deformities such as crossed beaks, humped backs, crooked feet, crooked keels, and wry tail; then double-check for disqualifications that are peculiar to your breed and variety. If you plan to exhibit Modern or Old English Game cocks, they must be dubbed and cropped (see page 256) to qualify for most shows. If you're among the many exhibitors who don't believe in dubbing and cropping, or who live in a state where doing so is illegal, show only cockerels, pullets, and hens.

No less than two months before the show, move each bird you wish to exhibit to its own small house with a grassy run. Although chickens need some sunshine for good health, show birds should not spend hour after hour in direct sunlight. Sunshine can fade the plumage of solid red breeds and those with red backgrounds, such as Mille Fleur bantams, or cause brassiness (yellowish metallic hue) in varieties with pale plumage — especially white, blue, or buff.

While you're moving the birds to individual housing, treat each one for lice and remove broken or off-color body feathers, giving the feathers plenty of time

Typical disqualifications: side sprig (left) in which the comb has an extra appendage, and wry tail (right) in which the tail feathers lean to one side.

to grow back. Unfortunately, incorrect coloring often shows up in flight feathers of the wing or sickle feathers of the tail, and these take a long time to grow back. To avoid injuring the feather follicles, never yank out a wing or tail feather. Instead, cut off the broken feather, leaving a 2-inch (5 cm) stub, then cut the shaft down the middle toward the skin. After a few days the feather will loosen and you can easily pull it out.

Coop Training

Rare is the bird that naturally shows well. Most exhibition birds need to be coop trained. The purpose of coop training is to let a bird know in advance what to expect and what's expected of it. Training minimizes stress, reducing the chance that the bird will experience health problems as a result of being shown. Training also lets a bird show to best advantage by getting it used to being handled and to being around people, and it ensures that the bird won't be disoriented by unfamiliar housing. An untrained bird can't show well if it walks or stands awkwardly because the coop floor is unfamiliar, crouches or flutters about because it's frightened by all the commotion, or struggles when handled by the judge.

No less than one week before the show, put each bird in a cage, or coop, similar to one that will be used at the show. Serious exhibitors purchase show coops for training purposes. These coops are quite costly, but you might reduce the price by purchasing used coops from a fair or club that's upgrading, or by combining your order with that of others in a group. Sources for show coops are listed in the appendix.

While the bird is in the coop, work with it two or three times a day. Keep each session short, so the bird doesn't become fatigued and lose interest. Begin by approaching the coop slowly. Avoid jerky movements, which can frighten even a well-trained bird. If the chicken appears the slightest bit frightened, stand quietly while it calms down. Begin your training session only when the bird appears calm.

First teach the bird to strike a pose suitable for its breed. Some breeds (Modern Game being the extreme) should take on an upright or vertical stance; others (such as Cochin) show best in a compact or horizontal stance. Birds can be taught to "set up" properly by tapping them in appropriate places with a judge's stick (a telescoping pointer available at office supply stores). To get a bird to lower its tail, for example, tap it above the tail. To raise the tail, tap below the tail. Teach the bird to stretch upward or bend downward, as appropriate, by getting it to reach for a bit of meat at the end of the judge's stick. Old-timers used raw hamburger, which is no longer suitable due to the danger of Salmonella poisoning. Instead, use tiny morsels of canned cat food. The meat not only serves as a tasty treat but also gives the bird a protein boost at a time of stress.

After spending a few minutes developing the bird's pose, next work on getting it used to being handled. Calmly open the door and maneuver the bird until it's standing sideways, with one wing facing you. Reach across the bird's back and place one hand over its far wing at the shoulder. Get a firm grip to keep the wing from flapping, and rotate the bird to face you.

Place your other hand under the bird's breast, with one of its legs between your thumb and index finger and the other leg between your second and third finger. Your index finger and second finger should be between the birds legs; the bird's keel should rest against your palm.

Gently lift the bird out of the coop head first. Always remove and replace the bird head first, so it can't catch its wings in the doorway and damage the feathers. With the bird outside the coop, hold it quietly for a moment, then remove your hand from its wing and let the bird sit in your hand another moment. At this point, you should be in full control. Imitate a judge's actions by turning the bird to examine its comb and wattles, and by opening each wing to examine the long feathers.

Gently return the bird to its coop, head first. If you're training a Cornish or other heavy breed, teach it to expect a judge to determine its body balance by dropping the bird 6 inches (15 cm) to the coop floor — a quick recovery

Minimize stress by letting each bird get used to the type of coop in which it will be housed at the show.

indicates good balance and leg placement. For other breeds, gently place the chicken on its feet and let go. A well-trained bird will continue to stand quietly.

The younger a bird is when you start, the more quickly it responds to coop training. But occasionally you'll run across one that does not care to be handled, no matter how patient you are or how much time you take. When a bird does not take kindly to being cooped at home, taking it to a show is a waste of time, energy, and money. If you have the patience, go ahead and work with the bird until the next show. Otherwise, concentrate your efforts on a more tractable individual.

Feeding

Sooner or later, all experienced exhibitors develop a custom diet for their show birds, taking into consideration (among other things) the desired weight for their breed and the effects of certain feeds on plumage color. A good starting place is basic breeder ration.

Do not feed yellow corn, especially to white varieties or those with white ear lobes. Corn tends to run to fat, and its pigment gives white birds a brassy hue. Feed no additional grain other than whole oats, offered free choice in a hopper separate from the breeder ration. Whole oats improve feather quality without making a bird fat. Be sure to offer granite grit as well, so the oats can be digested.

To stimulate natural oil production and give your bird a radiant glow, or bloom, include a small amount of oil-rich feed, such as safflower seed, sunflower seed, flax, or linseed meal. To further enhance the gloss of varieties with black or red plumage, feed a tiny amount of canned cat food daily, preferably offered as a treat during coop training.

Washing

A bird's plumage color can be enhanced, and the fluffiness of loosely feathered breeds emphasized, by washing. Light-colored birds should always be washed before a show. Dark-colored tightly feathered varieties need washing only if their plumage is heavily soiled.

Wash a chicken no less than 48 hours before a show, giving feathers time to get back some of their natural oil and giving you time to fix such goofs as streaking or improper shaping. I like to wash chickens in the laundry room. It's a warm, draft-free place to work and has a deep basin with plenty of warm running water. Others prefer to work in the garage or, if the weather is nice, outdoors, using three tubs filled with clean water — one for washing, two for rinsing.

If possible, use soft water, which cleans birds better than hard water. Begin with a basin full of warm water (90°F, 32°C). The temperature is right if you can hold your elbow in it without discomfort. Water that's too hot can cause a bird to faint. If a bird should faint, revive it by pouring a little cold water over its head.

Add enough shampoo or mild liquid dish soap to whip up a good head of suds. Don't use a harsh detergent, which makes feathers brittle. I've had great results with flea and tick shampoo for dogs and cats, which not only gets birds shiny clean but also zaps any lice or mites that may have gone unnoticed.

Place one hand against each of the chicken's wings so it can't flap (otherwise *you'll* get the bath), and slowly immerse the bird to its neck. If it struggles, dip its head briefly under water. Most birds relax as soon as they get the idea that they're in for a soothing bubble bath.

Thoroughly soak the bird by raising and lowering it, and drawing it back and forth through the water. With a sponge, soak the feathers through to the skin. To avoid breaking feathers, work only in the direction they grow. Rub in extra lather around the tail, where feathers tend to be stained by the oil gland.

When the chicken is thoroughly clean, lift it from the bath and press soapy water out with your hands, working from head to tail. If you're washing a crested bird, hold it upside down by the legs and dip the crest into the soapy water, keeping the bird's beak and eyes above water. Work suds into the topknot, or if the crest feathers are particularly dirty, apply a drop of shampoo directly to the topknot. Rinse the crest under running water to remove all traces of soap, taking care not to get any into the bird's eyes or nostrils.

Rinse the whole bird in fresh warm water that's slightly cooler than the wash water. Let the bird soak for a few minutes, then move it back and forth in the water to work out remaining soap. Lift the chicken from the rinse and press out excess water.

If any soap remains, the feathers will look dull and faded when they dry, so rinse the bird once more. This time add a little vinegar or lemon juice to remove the last vestiges of oil that may still be in the feathers.

If you're washing a white bird, brighten its plumage by adding two drops of liquid laundry bluing to this final rinse. If liquid bluing is no longer available in your part of the country, look for it in grocery and discount stores when you travel. To prevent streaking, thoroughly stir the bluing into the water. And remember, more is not better — add more than two drops and your white bird will turn blue.

Squeeze out excess water from the feathers and gently towel the bird off. Wrap the bird in a fresh towel and blot it to soak up remaining water.

Now that the bird's shanks and feet have been soaked, dirt and scales will be soft and easy to clean. Even if you don't wash the whole bird, at least soak its shanks and feet in warm water. Leg scales molt annually, just as feathers do, and you can easily remove dead semitransparent brittle scales by popping them off with a nail file, toothpick, or thumbnail. If necessary, use a toothpick to gently and carefully remove dirt beneath the scales and toenails. With a toothbrush and soapy water, scrub the shanks and toes. Use a little scouring powder on rough spots, but take care not to cause bleeding by rubbing too hard.

Grooming

To groom a bird after giving it a bath, tightly wrap it in a dry towel with only its head showing. Cover the feet with one corner of the towel so you can open the flap to work on its feet and legs.

Clean the comb and wattles with a little rubbing alcohol mixed with an equal amount of water, being careful not to get any into the bird's eyes. Dry the comb and rub it with baby oil, Vitamin E oil, or a mixture of equal parts alcohol and olive oil (but *not* petroleum jelly, which gets feathers greasy and gathers dust). Buff the comb until all the oil has been worked in, taking great care not to get oil on any feathers.

If you're grooming a breed with white ear lobes, such as Leghorn, Minorca, or Rosecomb, coat the washed and dried lobes with baby powder to keep them clean. Use a toothpick to clean around the bird's nostrils. Since a bird that spends much of its time caged can't keep its beak properly trimmed by scraping it on the ground, trim back the upper beak if necessary.

Now turn your attention to the legs and feet. Trim long nails with clippers or nail scissors. Excessively long toenails — usually occurring in caged birds that can't wear down their nails by walking on the ground — need to be trimmed in several stages, allowing a few days between each trimming for the quick to recede. If you wait until the last minute before a show, you'll run out of trimming time. Rub the cleaned feet and shanks with the same oil you used on the comb, again taking care not to get any on the feathers.

Release the cleaned, groomed bird into a clean training coop or pet carrier strewn with fresh litter. Any dirt that touches damp feathers may stick. Put only one bird in each coop to keep feathers from getting damaged or soiled.

You can dry birds outdoors, provided the temperature is at least 70°F (21°C) and either there's no breeze or you can provide good wind protection. If you're grooming loosely feathered birds like Cochins or Silkies, you can speed things up, as well as nicely puff out the feathers, with a hair dryer. A hair dryer also works well on crests.

Most breeds, though, look better if they dry slowly. Fluffing plumage with a hair dryer can be downright disastrous in tight-feathered breeds such as Cornish or Old English Game. If you must speed things up — or if you fear cold weather may chill your damp birds — put each one in a cardboard box and hang a heat lamp no closer than 2 feet (60 cm) above the bird. Or place individually caged birds in a room warmed to no more than 90°F (32°C), keeping them well away from the heater so they don't dry too quickly.

As the birds dry, arrange their feathers for proper shaping, especially around the base of the tail. Complete drying takes 12 to 18 hours, depending on the density of the feathers. To further whiten a white bird and give it a nice sheen, sponge off the plumage with hydrogen peroxide when the bird is half dry. When the white bird is fully dry, dust it lightly with cornstarch to keep it clean.

Faking

There's arguably a fine line between grooming and faking. Grooming is the process of making a bird look its best. Faking goes beyond grooming to alter a bird's appearance for the purpose of hiding natural defects. Arranging feathers while they dry is grooming; bending, breaking, or crimping feathers to change their natural angle is faking.

Faking has two purposes: to fool judges and to fool potential buyers. No one knows how common the practice is, since expert jobs are hard to detect and even harder to prove. The deceitful practice evolved into a high art in the days of the old-time stringmen, who traveled the show circuit exhibiting and selling their "string" of birds far from home.

Even today someone at a show may pull you aside to reveal "tricks of the trade," which are not only unethical but will get your bird disqualified if you have the bad luck to show under a judge who's knowledgeable enough to spot the fake job. Many judges, however, are reluctant to disqualify a bird that shows signs of having been tampered with, mainly because faking is so hard to prove. A cock, for example, may have a scarred comb because it got into a fight, or because a side sprig was snipped off or the comb was otherwise reshaped.

One faking practice that's both common and impossible to detect (unless you see it done) is the removal of stubs, or downy feathers on the legs or between the toes of a clean-legged bird. Some clean-legged breeds, including Wyandottes, readily sprout stubs. I once saw a fine Cornish cock disqualified at a show while its well-known and embarrassed owner stood by swearing the bird had no sign of stubs the day before. That may very well have been true: oiling the legs prior to judging likely loosened the scales enough to let the stubs slip out. The ultimate irony is that if this fellow had surreptitiously plucked the stubs, his bird would not have been disqualified.

Other clear cases of faking include using chemical solutions of various sorts to loosen or tighten plumage; starching tail feathers to make them stand up better; stitching a wry tail to straighten it; applying a coloring substance to the beak or ear lobes (some substances, like lipstick and rouge, are easy to detect because they rub off on plumage); and rubbing a caustic chemical on undesirable white ear lobes to make them blister, scab over, and turn red.

Borderline cases of faking most often involve feather color. White is the most likely color to be faked, since white plumage commonly looks brassy. Brassiness can occur for environmental reasons (e.g., too much sunshine or pigmented feed) or can be hereditary. Altering feather color by means of "softening, deepening, intensifying, or otherwise changing the *natural* color" is faking. Bringing out natural color by washing a bird, controlling its diet, or keeping it out of the sun is not faking. Is rinsing a bird with bluing or rubbing it down with hydrogen peroxide faking? Not if your intent is to bring out the natural whiteness of its plumage. But bleaching a bird with harsh chemicals to whiten naturally brassy feathers is faking. Just as bleaching human hair causes the hair to become brittle, bleaching a bird's plumage makes its feathers brittle. One criterion used to determine faking in white plumage is feather brittleness.

Feather pulling and beak trimming are additional practices that can be either grooming or faking. If you pull an off-color feather in plenty of time for a new one to grow back, hoping its color will be correct, that's grooming. If you remove an off-color feather on the sly 20 minutes before the judging starts, that's faking. Trimming the upper beak of a bird that's been housed where it couldn't keep its own beak worn down is grooming; trimming a crossed beak to hide the genetic defect is faking.

A crossed beak is a genetic deformity; such a bird should be culled.

The intent of faking is to make a bird look like something it is not. An unscrupulous exhibitor uses faking, rather than selective breeding, to make his birds appear superior to others at the show. An unscrupulous seller uses faking to peddle genetically defective culls to unsuspecting buyers. Either way, faking is fraud.

Show Carriers

An important part of show preparation is choosing an appropriate carrier in which to bring your chicken to the show. The carrier should be:

A long upper beak may be due to lack of wear; such a beak should be trimmed.

- ◆ *clean* — disinfected between shows
- ◆ *safe* — no protruding wires or nails
- ◆ *suitable in size* — not so small that the bird's crest or comb rubs the top or that foot feathers rub both sides at once; not so large that the bird has room to launch into panicked flight
- ◆ *well ventilated* — more so if the carrier will travel inside a car rather than in an open windy pick-up bed; if you carry chickens in the back of a truck, either cover the bed with a camper top or place each carrier inside a large cardboard box to protect the bird from drafts.

To prevent dirtied or damaged feathers, put no more than one bird in a carrier, line the carrier bottom with clean litter, and don't use wire carriers for feather-legged and crested birds.

Paint your name prominently on each carrier so that, in the scramble to coop out when the show ends, no one will mistakenly grab your box. The more expensive your carrier is, the more important some form of prominent marking becomes.

At the Show

Your birds will show best, and remain healthiest, if you make every attempt to reduce the stress they naturally experience as a result of being transported and shown. The biggest stress-reducing measures you can take are to avoid drafts and long periods without water or feed. Bring along the water and feed your chickens are used to. At the show, a bird may not drink water that tastes strange or eat rations with unfamiliar texture or composition.

Electrolytes and vitamin/mineral supplements help reduce stress when offered to birds for several days before and after a show, but these supplements should not be used during the show. Their taste could cause a bird in unfamiliar surroundings to go off feed or water, increasing its stress level.

Bring along drinkers that can't be dumped. Save small food cans. In the side of each, near the top, punch a hole and thread a piece of wire through. Use the wire to tie the can inside the show cage when you coop your birds in. If waterers are already provided, make sure they can't be knocked over by a frightened bird. Stabilize a loose waterer with string, a paper clip, a rubber band, or a piece of wire.

Be prepared to water and feed your own birds throughout the show. Many shows arrange to have someone travel from coop to coop making sure all birds have feed and water. Sometimes, however, that person is not reliable or has too many birds to handle, or the birds themselves dump their waterers or feeders in the excitement.

Feeders and Waterers

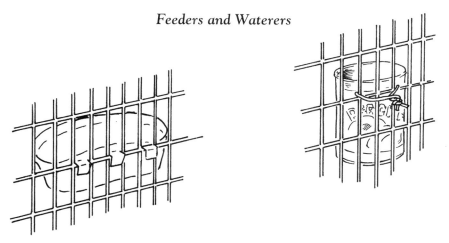

Whether store-bought galvanized or plastic (left) or homemade from a small can (right), cup feeders and waterers should be clipped securely to the show coop.

A few shows require individual exhibitors to be responsible for watering and feeding their own birds, as indicated in the premium list. Occasionally a show will not allow exhibitors to feed, water, or handle birds after they've been cooped in, since show officials have no way of knowing whether you are tending to your own birds or interfering with someone else's. The premium list should indicate a show's feed/water policy.

Last-Minute Touches

When you arrive at the show room, you will notice two types of exhibitors. One type will be running around tossing birds into their assigned coops. The other will be off in a corner, or sitting in the back of a truck or van, calmly examining, grooming, and reassuring each bird before placing it in its show coop. Keep an eye on those quiet fellows — they're your stiffest competition. They're also your best teachers.

Take a tip from them and give each of your chickens a last-minute going over. Polish the skin around the eyes, and the comb, wattles, ear lobes, legs, and toes of each bird with a touch of baby oil, mineral oil, Vitamin E oil, or a good quality hand lotion on a soft cloth. Use a mentholated rub on the combs and wattles to open up the chicken's airways and give your bird a healthy bloom. Be especially careful at this time not to get oil on plumage.

Arrange any feathers that may have gotten out of place during travel. To remove road dust and shine up the feathers, rub your *clean* bare hands or a piece of silk or wool over the plumage, working in the direction in which the feathers grow. If your bird is pitted against another bird of similar quality, the cleaner bird will invariably place higher.

Cooping In

In the showroom you'll be faced with row after row of show coops. "Cooping in" is the process of putting your bird in its assigned coop. The birds are organized according to classification, so that all the birds to be judged against one another are displayed in adjacent coops along one row (or in adjoining rows, if the class is large).

Attached to each coop is a coop card identifying both the bird assigned to that coop and the bird's owner. The card has room for the judge to note the bird's ranking (1, 2, 3, etc.). A thoughtful judge may also jot down some comment about the bird's outstanding good or bad feature (such as "nice lacing" or "too long in the back").

An open show is one in which the coop cards are left open during the judging. A closed show is one in which the cards are folded up so the judge can't see the name of the bird's owner and be influenced by the owner's reputation. Novices believe they have a better chance of winning at a closed show. Experienced exhibitors know that a skilled judge can identify birds coming from strains owned by top exhibitors simply by the birds' unique and uniform appearance. Some judges feel that closed shows insult their integrity.

And the Winner Is . . .

A licensed judge is supposed to appraise birds according to a scale of points published in the *Standard*. A specific number of points is allowed for each trait (comb, tail, back, symmetry, weight, condition, and so forth), with deductions (or cuts) made for such things as incorrect weight, missing tail feathers, off-color eyes, and other defects. In the old days, coop cards had this scale printed right on them, but rare is the judge today who refers to the point system at all. They say it takes too much time.

Judging these days tends to be much more subjective. Birds are often ranked according to the judge's personal likes and dislikes, which may diverge markedly from the *Standard*. In order to do well at a show, you have to know who the judge will be so you can show the kind of birds that judge likes to see in your breed and variety. As in all things, some judges are fairer than others. The fairest judges tend to be those who are willing to discuss the placings when the judging is over. The least fair judges don't like to explain their reasoning because they know they'll end up in an argument.

Differences in judging can be accounted for not only by differences in the judges' taste but also by differences in their conscientiousness, interpretation, powers of observation, experience, and age. Chickens, like other things in life, are subject to fads. Refinements in color or body type vary in desirability from year to year; what was "in" a generation ago may be "out" today, and vice versa. Successful exhibitors ignore these fads and work as closely as possible with the *Standard* descriptions for their breed and variety. (To avoid discouragement, however, keep in mind that the birds depicted and described in the *Standard* are the ideal to strive for, as determined by interested breeders and specialty clubs, and that such birds do not exist in real life.)

Chickens, like judges, also have their good days and their bad days. A bird may show differently from one day to the next due to changes in its condition, health, training, or stress level. Even the fairest of judges may place the same group of birds in one order on one day and in a different order on another day. If a bird doesn't place high in its class, don't be hasty to cull it until you find out why the bird did poorly. The best bird in the world won't win if it lacks maturity, is out of condition, or is out of sorts.

Ultimately, all judges base their decisions on two things:

- ◆ how each bird compares to the ideal (or standard) for its breed and variety;
- ◆ how each bird compares in type, condition, and training to the others in its class.

In ranking the birds in a class, most judges will first compare two birds, then compare the better of the two with a third bird, and so on until all the birds in the class have been ranked. A good way for novice exhibitors to become more observant is to do the same: rate each entry in the class according to its description in the *Standard*, then rank them against each other. When the judging is over, compare your ideas against the judge's. If your ranking is far off, ask a knowledgeable breeder to show you why. Regardless of your opinion, avoid getting into an argument with other exhibitors or with the judge. Even if every exhibitor in a class disagrees with the judge, the judge has the final say.

Clerking

An excellent way to learn the ins and outs of showing is to volunteer to help out. One of the best jobs is that of the judge's clerk, who gets to see firsthand how the judge determines the placing of each bird. If you're interested in a specific class, or in working with a specific judge (assuming the show is big enough to hire more than one), say so, but don't be upset if the assignment you want is already taken.

If you let the judge know ahead of time that you're there to learn, he or she may be willing to "think out loud" while judging. Avoid the temptation to pester the judge with questions, though. You'll only slow things down at a time when the judge needs to concentrate. One of the functions of a clerk, in fact, is to keep bystanders from talking to or questioning the judge.

To satisfy the crowd that invariably gathers around to watch, a good judge will discuss the placings after judging each class. A good clerk makes sure the judge has time for this important educational aspect of showing by keeping things running smoothly and efficiently. In doing so, the clerk locates each class to be judged, lets the judge know how many birds are in each class (so none will be missed if the class continues around the corner or across the aisle), keeps track of the judge's *Standard* and other paraphernalia that might otherwise be misplaced, records the judge's placings, and returns forms to the show secretary so ribbons and other awards can be distributed without delay.

Cooping Out

"Cooping out" refers to the removal of your bird from the showroom. Cooping out occurs in anticipation of going home, unless you are asked to coop a bird out early because it appears to be diseased. Having a bird disqualified is not sufficient reason for early coop out. Neither is selling a bird to someone who's anxious to take it home. Some shows have strong rules against people who coop out early, such as loss of eligibility to enter the next show. People who attend a poultry show come to see chickens, not empty cages.

So if you don't win, avoid the temptation to pick up your birds and go home. It's not sportsmanlike. Whether you win or lose, stick around after the judging to discuss the placings with other exhibitors. You'll be amazed at what you can learn.

The scheduled time for cooping out will be listed in the premium list. Coop-out time is usually a mad scramble, with everyone in a hurry to start for home. It's a good time to keep a close watch on your birds, carriers, and other equipment. If you have more stuff than you can carry outside in one trip, try to buddy up with someone so one of you can watch things inside while the other carries things to the parking lot.

Theft is so uncommon at shows that the disappearance of a superior bird or an expensive carrier causes a big stir among exhibitors. Even ordinary birds and simple cardboard boxes are sometimes inadvertently grabbed in the rush, especially by novice exhibitors or those who show large numbers of birds and don't take time to count them at coop-out time. Although such losses are not likely to be economically important, they're disheartening if not downright inconvenient. So, in addition to guarding your own belongings, take care in the fracas not to pack up something that's not yours.

Showing and Health

Some shows require you to get a health certificate from a veterinarian before entering your chickens. Unfortunately, even a health certificate doesn't ensure that a bird is healthy. A chicken can be fully capable of spreading a disease without appearing sick to the vet who signs the certificate. Indeed, some vets handle certification this way: "Have you had any problems with this bird? No? Well, it looks fine to me. Here's your certificate. That'll be thirty-five dollars."

Your chickens are less likely to catch a disease at a show if you take every precaution to reduce stress. Putting medications into the birds' drinking water is definitely a bad idea. Medications may cause a bird to drink less than usual, increasing stress and the likelihood of disease. Besides, if you don't know what disease your chickens are exposed to, if any, how would you know which medication to use?

When chickens do catch something at a show, it is likely to be a cold. Colds spread among birds the same way they spread among people — by coughing and sneezing. As you coop in, check the entries in adjoining coops. If a bird doesn't look healthy, or coughs and sneezes, notify show officials so the bird can be removed.

Upon return from a show, clean and disinfect all carriers, waterers, feeders, and other equipment you used at the show. Isolate returning birds for at least two weeks and watch them for signs of disease. Feed the isolated birds *after* you attend to the needs of your other birds. Despite these necessary precautions, take heart from the fact that a chicken rarely picks up disease at a show, since most people who exhibit their birds take pride in keeping them healthy.

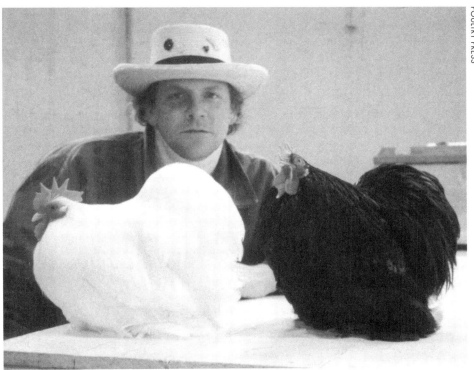

Tim Hooker of Illinois wins big with Cochin bantams — in a single show his white Cochin cockerel took grand champion and his black Cochin cockerel took reserve grand champion.

GLOSSARY

Abdominal capacity. Total abdominal depth and width.

Abdominal depth. The distance between the pubic bones and the keel bone.

Abdominal width. The distance between the two pubic bones.

Acute. Description of a disease with a severe and short development, often measured in hours and ending in death or recovery; opposite of chronic.

Alektorophobia. Fear of chickens.

American Standard of Perfection. A book published by the American Poultry Association describing each breed recognized by that organization.

Anthelmintic. An antiworm drug.

Avian. Pertaining to birds.

Bantam. A miniature chicken, about one-fourth the size of a regular-sized chicken.

Bantam Standard. A book published by the American Bantam Association describing each of the bantam breeds recognized by that organization.

Banty. Affectionate word for bantam.

Barny. Affectionate word for a barnyard chicken.

Barnyard chicken. A chicken of mixed breed.

Beak. The hard, protruding portion of a bird's mouth, consisting of an upper beak and a lower beak.

Beard. The feathers (always found in association with a muff) bunched under the beaks of such breeds as Antwerp Belgian, Faverolle, and Houdan.

Bedding. Straw, wood shavings, shredded paper, or anything else scattered on the floor of a chicken coop to absorb moisture and manure.

Biddy. Affectionate word for a hen.

Bill out. Use of the beak to scoop feed out of a trough onto the floor.

Biosecurity. Disease-prevention management.

Bleaching. The fading of color from the beak, shanks, and vent of a yellow-skinned laying hen.

Bloom. The moist, protective coating on a freshly laid egg that dries so fast you rarely see it; also, peak condition in an exhibition bird.

Blowout. Vent damage caused by laying an oversize egg.

Booted. Having feathers on the shanks and toes.

Break up. To discourage a hen from setting.

Breed. A group of chickens that are like each other and different from other chickens; also, pairing a rooster and hen for the purpose of obtaining fertile eggs.

Breeders. Mature chickens from which fertile eggs are collected; also, a person who manages such chickens.

Breed true. The characteristic of purebred chicks whereby they resemble both parents.

Broiler. A young, tender meat chicken; also called a "fryer."

Brood. To care for a batch of chicks; also, the chicks themselves.

Brooder. A mechanical device used to imitate the warmth and protection a mother hen gives her chicks.

Broody. A hen that covers eggs to warm and hatch them.

Cage fed. Description of chickens kept in cages.

Candle. To examine the contents of an intact egg with a light.

Candler. The light used to examine the contents of an intact egg.

Cannibalism. The bad habit chickens have of eating each other's flesh, feathers, or eggs.

Cape. The narrow feathers between a chicken's neck and back.

Carrier. An apparently healthy individual that transmits disease to other individuals; also, a container used to transport chickens.

Cecum. A blind pouch at the juncture of the small and large intestine (resembles the human appendix); plural: ceca.

Cestode. Tapeworm.

Chalazae. Two white cords on each side of a yolk that keep the yolk properly positioned within the egg white; singular: chalaza.

Check. An egg with a cracked shell, but with the shell membrane still intact.

Chronic. Description of a disease having long duration measured in days, months, or even years and being somewhat resistant to treatment; opposite of acute.

Class. A group of chickens competing against each other at a show.

Classification. The grouping of purebred chickens according to their place of origin, such as "American" or "Asiatic."

Clean legged. Having no feathers growing down the shanks.

Clinical. Having disease signs or symptoms that can be readily observed.

Cloaca. The chamber just inside the vent where the digestive, reproductive, and excretory tracts come together.

Cluck. The sound a hen makes to comfort her chicks; also, the hen herself.

Clucker. Affectionate word for a mother hen.

Clutch. A batch of eggs that are hatched together, either in a nest or in an incubator (from the Old Norse word "klekja," meaning *to hatch*), also called a "setting"; also, all the eggs laid by a hen on consecutive days, before she skips a day and starts a new laying cycle.

Coccidiasis. Infection with coccidial protozoa without showing any signs.

Coccidiosis. A parasitic protozoal infestation, usually occurring in damp, unclean housing.

Coccidiostat. A drug used to keep chickens from getting coccidiosis.

Cock. A male chicken; also called a "rooster."

Cockerel. A male chicken under 1 year old.

Condition. A chicken's state of health and cleanliness.

Conformation. A chicken's body structure.

Contagious. Description of a disease that's readily transmitted from one individual or flock to another.

Coop. The house or cage in which a chicken lives.

Crest. A puff of feathers on the heads of breeds such as Houdan, Silkie, or Polish; also called a "topknot."

Crop. A pouch at the base of a chicken's neck that bulges after the bird has eaten; also, to trim a bird's wattles.

Crossbreed. The offspring of a hen and a rooster of two different breeds.

Cull. To eliminate a non-productive or inferior chicken from a flock; also, the non-productive or inferior chicken itself.

Dam. Mother.

Dam family. Sibling chickens that all have the same dam as well as sire.

Debeak. To remove a portion of a bird's top beak to prevent cannibalism.

Defect. Any characteristic that makes a chicken less than perfect.

Depopulate. To get rid of an entire flock.

Dirties. Eggs with dirt or stains covering more than ¼ of the shell.

Disinfectant. Anything used to destroy disease-causing organisms.

Disqualification. A defect or deformity serious enough to bar a chicken from a show.

Dominecker/Dominicker. Colloquialism for Dominique, often erroneously applied to barred Plymouth Rocks.

Down. The soft, fur-like fluff covering a newly hatched chick; also, the fluffy part near the bottom of any feather.

Drench. To give liquid medication orally; also, the liquid medication itself.

Droppings. Chicken manure.

Dub. To trim the comb.

Dust. The habit chickens have of thrashing around in soft soil to clean their feathers and discourage body parasites.

Egg tooth. A horny cap on a chick's upper beak that helps the chick pip through the shell.

Electrolytes. Natural chemicals in the blood needed by body cells to maintain balance.

Embryo. A fertilized egg at any stage of development prior to hatching.

Enteritis. Inflammation of the intestine.

Eversion. Turned inside out.

Exhibition breeds. Chickens kept and shown for their beauty rather than their ability to lay eggs or produce meat.

Fake. The dishonest practice of concealing a defect or disqualification from a potential buyer or show judge.

Feather legged. Having feathers growing down the shanks.

Fecal. Pertaining to feces.

Feces. Droppings or body waste.

Fertile. Capable of producing a chick.

Fertilized. Containing sperm.

Finish. The amount of fat beneath the skin of a meat bird.

Flock. A group of chickens living together.

Forced-air incubator. A mechanical device for hatching fertile eggs that has a fan to circulate warm air.

Fowl. Domesticated birds raised for food; also, a stewing hen.

Free range. To allow chickens to roam pasture at will.

Frizzle. Feathers that curl rather than laying flat.

Fryer. A tender young meat chicken; also called a "broiler."

Gizzard. An organ that contains grit for grinding up the grain and plant fiber a chicken eats.

Go light. Grow thin while eating ravenously; synonym for anemia.

Grade. To sort eggs according to their interior and exterior qualities.

Grit. Sand and small pebbles eaten by a chicken and used by its gizzard to grind up grain and plant fiber.

Ground fed. Free to move about outdoors, as opposed to being housed and fed within a building or a cage.

Hackles. A rooster's cape feathers.

Hatch. The process by which a chick comes out of the egg; also, a group of chicks that come out of their shells at roughly the same time.

Hatchability. Percentage of fertilized eggs that hatch under incubation.

Hatching egg. A fertilized egg stored in a way that does not destroy its ability to hatch.

Helminth. A category of parasitic worms.

Helminthiasis. Parasitic worm infestation.

Hen. A female chicken.

Hen feathered. The characteristic of a rooster having rounded rather than pointed sex feathers.

Host. A bird (or other animal) on or in which a parasite or an infectious agent lives.

Hybrid. The offspring of a hen and rooster of different breeds, each of which might themselves be crossbred; often erroneously applied to the offspring of a hen and rooster of different strains within a breed.

Immunity. Ability to resist infection.

Impaction. Blockage of a body passage or cavity, such as the crop or cloaca.

Incubate. To maintain favorable conditions for hatching fertile eggs.

Incubation period. The time it takes for a bird's egg to hatch; also, the time it times from exposure to a disease-causing agent until the first symptom appears.

Incubator. A mechanical device for hatching fertile eggs.

Infectious. Capable of invading living tissue and multiplying therein, causing disease.

Infertility. Temporary or permanent inability to reproduce.

Intensity of lay. The number of eggs a hen lays during a given time.

Keel. The breastbone, which resembles the keel of a boat.

Leaker. An egg that leaks because the shell is cracked and the shell membrane is broken.

Litter. Bedding.

Mate. The pairing of a rooster with one or more hens; a hen or rooster so paired.

Mite. A tiny jointed-legged body parasite.

Molt. The annual shedding and renewing of a bird's feathers.

Morbidity. Percentage affected by a disease.

Mortality. Percentage killed by a disease.

Muff. The feathers (always found in association with a beard) sticking out from both sides of the face of such breeds as Antwerp Belgian, Faverolle, and Houdan; also, called "whiskers."

Nematode. Roundworm.

Nest. A secluded place where a hen feels she may safely leave her eggs; also, the act of brooding.

Nest egg. A wooden or plastic egg placed in a nest to encourage hens to lay there.

Nest run. Ungraded eggs.

Oocyst. The infective fertilized egg of certain one-celled animal parasites including protozoa.

Oviduct. The tube inside a hen through which an egg travels when it is ready to be laid.

Oviposition. The laying of an egg.

Parasite. An organism that lives on or inside a host animal and derives food or protection from the host without giving anything in return.

Pasting. Loose droppings sticking to vent area.

Pathogenic. Capable of causing disease.

Pathologist. A medical professional who examines internal damage caused by disease.

Peck order. The social rank of chickens.

Pen. A group of chickens entered into a show and judged together; also, a group of chickens housed together for breeding purposes.

Perch. The place where chickens sleep at night; the act of resting on a perch; also called "roost."

Persistency of lay. The ability of a hen to lay steadily over a long period of time.

Pickout. Vent damage due to cannibalism.

Pigmentation. The color of a chicken's beak, shanks, and vent.

Pinbones. Pubic bones.

Pinfeathers. The tips of newly emerging feathers.

Pip. The hole a newly formed chick makes in its shell when it is ready to hatch; also, the act of making the hole.

Plumage. The total set of feathers covering a chicken.

Postmortem. Pertaining to or occurring after death.

Poultry. Chickens and other domesticated birds raised for food.

Predator. One animal that hunts another for food.

Processor. A person or firm that kills, cleans, and packages meat birds.

Producer. A person or firm that raises meat birds or layers.

Prolapse. Slipping of a body part from its normal position, often erroneously used to describe an everted organ.

Protozoan. A single-celled microscopic animal that may be either parasitic or beneficial; plural: protozoa.

Proventriculus. A chicken's stomach.

Pubic bones. Two sharp, slender bones that end in front of the vent; also called "pinbones."

Pullet. A female chicken under 1 year old.

Purebred. The offspring of a hen and rooster of the same breed.

Rales. Any abnormal sound coming from the airways.

Range fed. Description of chickens that are allowed to graze pasture.

Ration. The combination of all feed consumed in a day.

Reportable disease. A disease so serious it must, by law, be reported to a state or federal veterinarian.

Resistance. Immunity to infection.

Restricted egg. A check, dirty, leaker, or otherwise inedible egg.

Roaster. A cockerel or pullet, usually weighing 4 to 6 pounds, suitable for cooking whole in the oven.

Roost. The place where chickens spend the night; the act of resting on a roost; also called "perch."

Rooster. A male chicken; also called a "cock."

Saddle. The part of a chicken's back just before the tail.

Scales. The small, hard, overlapping plates covering a chicken's shanks and toes.

Scratch. The habit chickens have of scraping their claws against the ground to dig up tasty things to eat; also, any grain fed to chickens.

Scratcher. Affectionate word for a range-fed chicken.

Set. To keep eggs warm so they will hatch; also called "brood."

Setting. A group of hatching eggs in an incubator or under a hen; the incubation of eggs by a hen (incorrectly called "sitting" by those who try too hard to be grammatically correct).

Sexed. Newly hatched chicks that have been sorted into pullets and cockerels.

Sex feather. A hackle, saddle, or tail feather that is rounded in a hen but usually pointed in a rooster (except in breeds that are hen feathered).

Shank. The part of a chicken's leg between the claw and the first joint.

Sickles. The long, curved tail feathers of some roosters.

Sire. Father.

Sire family. The offspring of one cock mated to two or more hens, so that all are full or half siblings.

Smut. Black feathers that are uncharacteristic for the breed, such as black body feathers in a Rhode Island Red.

Spent. No longer laying well.

Sport. Genetic mutation that occurs naturally or is induced (for example, by radiation); also, cockfighting.

Spurs. The sharp pointed protrusions on a rooster's shanks.

Stag. A cockerel on the brink of sexual maturity, when his comb and spurs begin to develop.

Standard. The description of an ideal specimen for its breed; also, a chicken that conforms to the description of its breed in the *American Standard of Perfection*, somtimes erroneously used when referring to large as opposed to bantam breeds.

Started pullets. Young female chickens that are nearly old enough to lay.

Starter. A ration for newly hatched chicks.

Starve-out. Failure of chicks to eat.

Steal. A hen's instinctive habit of hiding her eggs.

Sterile. Permanent inability to reproduce.

Sternum. Breastbone or keel.

Still-air incubator. A mechanical device for hatching fertile eggs that does not have a fan to circulate air.

Straightbred. Purebred.

Straight run. Newly hatched chicks that have not been sexed; also called "unsexed" or "as hatched."

Strain. A flock of related chickens selectively bred by one person or organization for so long that the offspring have become uniform in appearance or production.

Stress. Any physical or mental tension that reduces resistance.

Stub. Down on the shank or toe of a clean-legged chicken.

Tin hen. A metal incubator.

Trachea. Windpipe.

Trematode. A parasitic fluke.

Trio. A cock and two hens or a cockerel and two pullets of the same breed and variety.

Type. The size and shape of a chicken that tells you what breed it is.

Unthrifty. Unhealthy appearing and/or failing to grow at a normal rate.

Urates. Uric acid (salts found in urine).

Vaccine. Product made from disease-causing organisms and used to produce immunity.

Variety. Subdivision of a breed according to color, comb style, beard, or leg feathering.

Vent. The outside opening of the cloaca, through which a chicken emits eggs and droppings from separate channels.

Virulence. Strength of an organism's ability to cause disease.

Warfarin. An anticoagulant used to poison rodents.

Wattles. The two red or purplish flaps of flesh that dangle under a chicken's chin.

Whiskers. Muffs.

Zoning. Laws regulating or restricting the use of land for a particular purpose, such as raising chickens.

STATE RESOURCES

If you have a poultry-related question or problem, contact the poultry specialist at your state's land-grant college (listed below under each state). Consult a pathology or diagnostic laboratory *only* if your flock experiences a serious disease problem *and* you can't get veterinary help locally. Some path labs require either referral by a licensed veterinarian or a written statement indicating why a veterinary referral isn't possible (such as you live in a remote area where no vet is reasonably available). Be prepared to submit a few diseased and recently dead birds for examination.

Alabama

Extension Poultry Specialist, Poultry Science Department, Auburn University, Auburn, AL 36849, 334-844-2613

C.S. Roberts Veterinary Diagnostic Laboratory, PO Box 2209, Auburn, AL 36831, 334-844-4987

J.B. Taylor Veterinary Diagnostic Laboratory, 495 AL 203, Elba, AL 36323, 334-897-6340

Veterinary Diagnostic Laboratory, Boaz, AL 35957, 205-593-2995

Alaska

Alaska State Federal Laboratory, 500 South Alaska Street, Suite A, Palmer, AK 99645, 907-745-3236

Extension Livestock Specialist, PO Box 756180, University of Alaska, Fairbanks, AK 99775, 907-474-7083

Arizona

Extension Poultry Specialist, University of Arizona, Tucson, AZ 85721, 520-621-1980

Veterinary Diagnostic Laboratory, University of Arizona, Tucson, AZ 85721, 520-621-2356

Arkansas

Extension Poultry Specialist, Department of Agriculture, University of Arkansas at Pine Bluff, PO Box 82, Pine Bluff, 71601, 501-543-8526

Livestock and Poultry Commission Diagnostic Laboratory, PO Box 8505,
1 Natural Way Resource Drive, Little Rock, AR 72215, 501-225-5650
Livestock and Poultry Commission Diagnostic Laboratory, Highway 71 N, 3559
N Thompson, Springdale, AR 72764, 501-751-4869

California

California Veterinary Diagnostic Laboratory System, University of California,
PO Box 1770, Davis, CA 95617, 916-752-8700
Extension Poultry Specialist, Department of Avian Sciences, University of
California, Davis, CA 95616, 916-752-3513
Fresno Branch Laboratory, 2789 South Orange Avenue, Fresno, CA 93725,
209-498-7740
San Bernardino Branch Laboratory, 105 West Central Avenue, San Bernardino,
CA 92408, 909-383-4287
San Diego County Veterinary Laboratory, 5555 Overland Avenue, Building 4,
San Diego, CA 92123, 619-694-2838
Tulare Branch Laboratory, 18830 Road 112, Tulare, CA 93274, 209-688-7543
Turlock Branch Laboratory, PO Box 1522, Turlock, CA 95381, 209-634-5837

Colorado

CSU Veterinary Diagnostic Laboratory, 27847 RD 21, Rocky Ford, CO 81067,
303-254-6382
Extension Poultry Specialist, Department of Animal Sciences, Colorado State
University, Fort Collins, CO 80523, 970-491-7803
Veterinary Diagnostic Laboratory, Colorado State University, Fort Collins, CO
80523, 970-491-1281
Western Slope Animal Diagnostic Laboratory, 425 29 Road, Grand Junction,
CO 81501, 970-243-0673

Connecticut

Department of Pathobiology, Box U-89, 61 North Eagleville Road, University of
Connecticut, Storrs, CT 06269, 860-486-3736
Extension Poultry Specialist, Department of Animal Science, Box U-40, 3636
Horsebarn Road Ext., University of Connecticut, Storrs, CT 06268,
860-486-1008

Delaware

Extension Poultry Specialist, Poultry Research Laboratory, Route 2 Box 48,
Georgetown, DE 19947, 302-856-7303

Poultry and Animal Health Section, State Department of Agriculture, PO Drawer D, 2320 S. Dupont Highway, Dover, DE 19901, 302-739-4811

Florida

Extension Poultry Specialist, Poultry Science Department, PO Box 110920, University of Florida, Gainesville, FL 32611, 352-392-1931

Live Oak Diagnostic Laboratory, PO Box Drawer O, Live Oak, FL 32060, 904-362-1216

Georgia

Poultry Laboratory, 150 Tom Freyer Drive, Douglas Airport, Douglas, GA 31533, 912-384-3719

Poultry Laboratory, PO Box 20, Oakwood, GA 30566, 770-535-5996

Poultry Specialist, Extension Poultry Science Deparment, 4 Towers Building, University of Georgia, Athens, GA 30602, 706-542-1325

Hawaii

Extension Poultry Specialist, Department of Animal Sciences, 1800 East-West Road, University of Hawaii, Honolulu, HI 96822, 808-956-8334

Veterinary Laboratory Branch, Hawaii Department of Agriculture, 99-941 Halawa Valley Street, Aiea, HI 96701, 808-483-7100

Idaho

Extension Poultry Specialist, Southeast Idaho Research & Extension Center, 16952 S. Tenth Avenue, University of Idaho, Caldwell, ID 83605, 208-459-6365

Division of Animal Industries, PO Box 7249, 2230 Old Penitentiary Road, Boise, ID 83707, 208-332-8540

Illinois

Animal Disease Laboratory, 9732 Shattuc Road, Centralia, IL 62801, 618-532-6701

Animal Disease Laboratory, PO Box 2100X, 2100 South Lake Storey Road, Galesburg, IL 61401, 309-344-2451

Extension Poultry Specialist, Department of Animal Sciences, 132 Animal Science Laboratory, 1207 W. Gregory Drive, University of Illinois, Urbana, IL 61801, 217-244-0195

Indiana

Animal Disease Diagnostic Laboratory, Purdue University, 1175 ADDL, West
Lafayette, IN 47907, 765-494-7440
Animal Disease Diagnostic Laboratory, SIPAC, 11367 E. Purdue Farm Road,
Dubois, IN 47527, 812-678-3401
Extension Poultry Specialist, Department of Animal Sciences, Lilly Hall, Purdue
University, West Lafayette, IN 47907, 765-494-8009

Iowa

Extension Poultry Specialist, Department of Animal Sciences, Iowa State
University, Ames, IA 50011, 515-294-4303
Veterinary Diagnostic Laboratory, Iowa State University, Ames, IA 50011,
515-294-1950

Kansas

Extension Poultry Specialist, Department of Animal Sciences, Call Hall, Kansas
State University, Manhattan, KS 66506, 913-532-6533
Diagnostic Medicine Pathology Laboratories, College of Veterinary Medicine,
Veterinary Medical Center, Manhattan, KS 66506, 913-532-5650

Kentucky

Extension Poultry Specialist, Department of Animal Sciences, Garrigus Building,
University of Kentucky, Lexington, KY 40546, 606-257-7529
Livestock Disease Diagnostic Center, 1429 Newtown Pike, Lexington, KY
40511, 606-253-0571
Murray State University Veterinary Center, Breathitt Veterinary Center, PO
Box 2000, 715 North Drive, Hopkinsville, KY 42240, 502-886-3959

Louisiana

Central Louisiana Livestock Diagnostic Laboratory, 217 Middleton Drive,
Lecopte, LA 71346, 318-473-6500
Extension Poultry Specialist, Department of Poultry Science, Louisiana State
University, Baton Rouge, LA 70803, 504-388-4481
Louisiana Veterinary Medical Diagnostic Laboratory, LSU School of Veterinary
Medicine, South Stadium Drive, Baton Rouge, LA 70803, 504-346-3193

Maine

Extension Poultry Specialist, Animal & Veterinary Sciences, 127 Hitchner Hall,
University of Maine, Orono, ME 04469, 207-581-2768

Pathology Diagnostic and Research Laboratory, 105 Hitchner Hall, University of
Maine, Orono, ME 04469, 207-581-2771

Maryland

Animal Health Department Laboratory, PO Box 376, Oakland, MD 21550,
501-334-2185
Animal Health Laboratory, Maryland Department of Agriculture, 211 Safety
Drive, Centreville, MD 21617, 401-758-0846
Animal Health Laboratory, Maryland Department of Agriculture, 1840
Rosemont Avenue, Frederick, MD 21702, 301-663-9528
Animal Health Laboratory, Maryland Department of Agriculture, PO Box J,
Salisbury, MD 21802, 410-543-6610
Animal Health Laboratory-Central, Greenmead Drive, College Park, MD 20740,
301-935-6074
Extension Poultry Specialist, Poultry Research & Education Facility, 11990
Strickland Drive, Princess Anne, MD 21853, 410-651-9111

Massachusetts

Avian Diagnostic Laboratory, University of Massachusetts, Suburban Experi-
ment Station, 240 Beaver Street, Waltham, MA 02154, 617-891-0650
Extension Poultry Specialist, Department of Veterinary & Animal Sciences,
Paige Laboratory, University of Massachusetts, Amherst, MA 01003,
413-545-2312

Michigan

Animal Health Diagnostic Laboratory, PO Box 30076, Lansing, MI 48909,
517-353-1683
Extension Poultry Specialist, Animal Science Department, 102 Anthony Hall,
Michigan State University, East Lansing, MI 48824, 517-353-2906

Minnesota

Extension Poultry Specialist, Department of Avian Sciences, 120 Peters Hall,
University of Minnesota, St. Paul, MN 55108, 612-624-4928
Minnesota Veterinary Diagnostic Laboratories, 1333 Gotonor Avenue, College
of Veterinary Medicine, University of Minnesota, St. Paul, MN 55108,
612-625-8787

Mississippi

Extension Poultry Specialist, Poultry Science Department, PO Box 5188,
Mississippi State University, Mississippi State, MS 39762, 601-325-3416

Mississippi Veterinary Diagnostic Laboratory, PO Box 4389, 2531 N. West Street, Jackson, MS 39216, 601-354-6091

Missouri

Extension Poultry Specialist, Animal Sciences Department, S105 Animal Science Center, University of Missouri, Columbia, MO 65211, 573-882-6658

Veterinary Diagnostic Laboratory, Missouri Department of Agriculture, 1922 North Broadway, Springfield, MO 65803, 471-865-2261

Veterinary Medical Diagnostic Laboratory, College of Veterinary Medicine, PO Box 6023, University of Missouri, Columbia, MO 65205, 573-882-6811

Montana

Extension Poultry Specialist, 418 Mineral Avenue, Montana State University, Libby, MT 59923, 406-293-7781

Veterinary Diagnostic Laboratory Division, Montana Department of Livestock, PO Box 997, 19th & Lincoln Streets, Bozeman, MT 59771, 406-994-4885

Nebraska

Diagnostic Laboratory, Department of Veterinary Science, University of Nebraska, Lincoln, NE 68583, 402-472-1434

Extension Poultry Specialist, Animal Science Department, PO Box 830908, University of Nebraska, Lincoln, NE 68583, 402-472-6451

University of Nebraska, West Central Research and Extension Center, Box 46A Rt. 4, North Platte, NE 69101, 308-532-3611

Nevada

Animal Disease Laboratory, Nevada Department of Agriculture, PO Box 11100, 350 Capitol Hill Avenue, Reno, NV 89502, 702-789-0185

Extension Livestock Specialist, University of Nevada, Reno, NV 89557, 702-397-2184

New Hampshire

Extension Poultry Specialist, Department Animal & Nutritional Sciences, Kendall Hall, University of New Hampshire, Durham, NH 03824, 603-862-2247

Veterinary Diagnostic Laboratory, University of New Hampshire, 319 Kendall Hall, Durham, NH 03824, 603-862-2726

New Jersey

Extension Poultry Specialist, Animal Science Department, Bartlett Hall, Cooks
College--Rutgers State University, New Brunswick, NJ 08903,
732-932-9793
New Jersey Animal Health Diagnostic Laboratory, John Fitch Plaza PO Box 330,
Trenton, NJ 08625, 609-292-3965

New Mexico

Extension Poultry Specialist, Department Animal & Range Science, PO Box
30003, New Mexico State University, Las Cruces, NM 88003,
505-646-3016
Veterinary Diagnostic Services, 700 Camino De Salud, NE, Albuquerque, NM
87106, 505-841-2576

New York

Cornell University Duck Research Laboratory, Box 217, 192 Old Country Road,
Eastport, NY 11941, 516-325-0600
Department of Microbiology and Immunology, New York State College of
Veterinary Medicine, Ithaca, NY 14853, 607-253-3365
Extension Poultry Specialist, Animal Science Department, Morrison Hall,
Cornell University, Ithaca, NY 14853, 607-255-8143

North Carolina

Extension Poultry Specialist, Department of Poultry Science, PO Box 7608,
North Carolina State University, Raleigh, NC 27695, 919-515-5391
Rollins Animal Disease Diagnostic Laboratory, PO Box 12223 Cameron Village
Station, 2101 Blue Ridge Road, Raleigh, NC 27605, 919-733-3986

North Dakota

Extension Poultry Specialist, Animal & Range Science, Hultz Hall, North
Dakota State University, Fargo, ND 58105, 701-237-7691
North Dakota Veterinary Diagnostic Laboratory, North Dakota State University
Fargo, ND 58102, 701-231-7551

Ohio

Animal Disease Diagnostic Laboratory, Ohio Department of Agriculture,
8995 E. Main Street, Reynoldsburg, OH 43068, 614-728-6220
Extension Poultry Specialist, Animal Sciences Department, Plumb Hall,
2027 Ohio State University, Columbus, OH 43210, 614-728-6220

Oklahoma

Extension Poultry Specialist, Animal Science Building, Room 201, Oklahoma State University, Stillwater, OK 74078, 405-744-9293
Oklahoma Animal Disease Diagnostic Laboratory, College of Veterinary Medicine, Oklahoma State University, Stillwater, OK 74078, 405-744-6623

Oregon

Extension Poultry Specialist, Poultry Science Department, Withy Combe Hall, Oregon State University, Corvallis, OR 97331, 541-737-2254
Oregon State Department of Agriculture, Animal Health Laboratory, 635 Capitol Street NE, Salem, OR 97310, 503-986-4550
Veterinary Diagnostic Laboratory, Oregon State University, PO Box 429, Corvallis, OR 97339, 541-737-3261

Pennsylvania

State Veterinary Laboratory, Pennsylvania Department of Agriculture, 2305 North Cameron Street, Harrisburg, PA 17110, 717-787-8808
Extension Poultry Specialist, 1383 Arcadia Road Room 1, Lancaster, PA 17601, 717-394-6851
Poultry Diagnostic Laboratory, Veterinary Science Department, Pennsylvania State University, University Park, PA 16802, 814-863-0839
Poultry Diagnostic Laboratory, Delaware Valley College of Science & Agriculture, Doylestown, PA 18901, 215-345-1500
Regional Diagnostic Laboratory, 5349 William Flynn Highway Route 8, Gibsonia, PA 15044, 412-443-1585

Puerto Rico

Department of Agriculture Veterinary Laboratory, PO Box 490, Dorado, PR 00646, 809-796-1650
Extension Poultry Specialist, Animal Industry Department, University of Puerto Rico, Mayaguez, PR 00708, 809-832-4040

Rhode Island

Department of Fisheries, Animals, and Veterinary Science, University of Rhode Island, Kingston, RI 02881, 401-874-2477
Extension Poultry Specialist, Department of Fisheries, Animals & Veterinary Sciences, Peckham Farm, University of Rhode Island, Kingston, RI 02881, 401-874-2072

South Carolina

Clemson Animal Diagnostic Laboratory, PO Box 102406, Columbia, SC 29224, 864-788-2260
Extension Poultry Specialist, Poultry Science Department, 129 P&AS Building, Clemson University, Clemson, SC 29634, 864-656-4026

South Dakota

Animal Disease Research & Diagnostic Laboratory, South Dakota State University, 105 Veterinary Science, PO Box 2175, Brookings, SD 57007, 605-688-5171
Poultry Specialist, Animal Science Complex, PO Box 2170, South Dakota State University, Brookings, SD 57007, 605-693-3484

Tennessee

C.E. Kord Animal Disease Laboratory, Ellington Agriculture Center Porter Building, PO Box 40627 Melrose Station, Nashville, TN 37204, 615-837-5125
Extension Poultry Specialist, Animal Science Department, PO Box 1071, University of Tennessee, Knoxville, TN 37901, 423-974-7351

Texas

Extension Poultry Specialist, Poultry Science Department 2472, Texas A&M University, College Station, TX 77843, 409-845-4318
TVMDL Poultry Diagnostic Lab, PO Box 187, Center, TX 75935, 409-598-4451
Department of Pathology and Biology, College of Veterinary Medicine, Texas A&M University, College Station, TX 77843, 409-845-5941
Poultry Disease Laboratory, Texas Veterinary Medical Diagnostic Laboratory System, PO Box 84, 1812 Water Street, Gonzales, TX 78629, 830-672-2834

Utah

Extension Poultry Specialist, Animal & Veterinary Sciences, Utah State University, Logan, UT 84322, 801-797-2145
Department of Agriculture, State Chemist Office, 350 North Redwood Road, Salt Lake City, UT 84114, 801-538-7128
Utah State University-Provo Veterinary Laboratory, Utah Agricultural Experimental Station, 2031 S. State, Provo, UT 85606, 801-373-6383
Veterinary Diagnostic Laboratory, Utah State University, Logan, UT 84322, 801-750-1895

Vermont

Extension Poultry Specialist, Department of Animal Sciences, Carrigan Hall, University of Vermont, Burlington, VT 05405, 802-656-2074

Livestock Division Laboratory, Vermont Department of Agriculture, 103 South Main Street, Waterbury, VT 05671, 802-828-2412

Virginia

Division of Animal Health, Regulatory Laboratory, 116 Reservoir Street, Harrisonburg, VA 22801, 540-434-3897

Division of Animal Health, Regulatory Laboratory, Route 460 34591 General Mahone Blvd., Ivor, VA 23866, 757-859-6221

Division of Animal Health, Regulatory Laboratory, 272 Academy Hill Road, Warrenton, VA 20186, 540-347-6385

Division of Animal Health, Regulatory Laboratory, Box 738 Cassel Road, Wytheville, VA 24382, 540-228-5501

Extension Poultry Specialist, Department of Animal & Poultry Sciences, Virginia Polytechnic Institute, Blacksburg, VA 24061, 540-231-5087

Veterinary Teaching Hospital, VA-MD Regional College of Veterinary Medicine, Virginia Polytechnic Institute & State University, Blacksburg, VA 24061, 540-961-4621

Virginia Department of Agriculture & Consumer Services, Division of Animal Health Regulatory Laboratory, Bureau of Laboratory Services, 4832 Tyreanna, Lynchburg, VA 24504, 804-947-6731

Washington

Animal Disease Diagnostic Laboratory, PO Box 2037 College Station, Washington State University, Pullman, WA 99165, 509-335-9696

Extension Poultry Specialist, Animal Sciences Department, Puyallup Research and Extension Center, 7612 Pioneer Way East, Puyallup, WA 98371, 253-445-4579

Poultry Diagnostic Laboratory, Washington State University, 7612 Pioneer Way East, Puyallup, WA 98371, 253-445-4536

West Virginia

Extension Poultry Specialist, Division of Animal & Veterinary Science, Agricultural Sciences Building, University of West Virginia, Morgantown, WV 26506, 304-293-2406

State-Federal Laboratory, West Virginia Department of Agriculture, Capitol Building, Charleston, WV 25305, 304-348-3418

West Virginia Regional Animal Health Laboratory, Route 1 Box 302, Moorefield, WV 26836, 304-538-2397

Wisconsin

Extension Poultry Specialist, Poultry Science Department, 260 Animal Sciences, 1675 Observatory Drive, University of Wisconsin, Madison, WI 53706, 608-262-9764

Regional Animal Health Laboratory, 1521 East Guy Avenue, Barron, WI 54812, 715-537-3151

Wisconsin Animal Health Laboratory, 6101 Mineral Point Road, Madison, WI 53705, 608-266-2465

Wyoming

Extension Poultry Specialist, University of Wyoming, Laramie, WY 82071, 307-766-3100

Wyoming State Veterinary Laboratory, 1174 Snowy Range Road, Laramie, WY 82070, 307-742-6638

Canada

Alberta

Poultry Pathology, 3438 O.S. Longman Building, 6909-116 Street, Edmonton, AB T6H 4P2, 403-427-2238

Poultry Specialist, Department of Animal Science, 310 Agriculture/Forestry Building, University of Alberta, Edmonton, AB T6G 2P5, 403-492-3234

British Columbia

Poultry Specialist, Animal Science Department, University of British Columbia, Vancouver, BC V6T 1Z4, 604-822-2355

Manitoba

Poultry Specialist, Department of Animal Science, University of Manitoba, Winnipeg, MB R3T 2N2, 204-474-9383

Provincial Veterinary Laboratory, Agricultural Services Complex, University of Manitoba, 545 University Crescent, Winnipeg, MB R3T 2N2, 204-945-7650

New Brunswick

Poultry Specialist, New Brunswick Agriculture, Box 5001, Sussex, NB E0E 1P0
Veterinary Pathology, Department of Agriculture, PO Box 6000, Fredericton,
NB E3B 5H1, 506-453-2210

Nova Scotia

Poultry Specialist, Department of Animal Science, Nova Scotia Agricultural
College, Truro, NS B2N 5E3, 902-893-6644
Veterinary Pathology Laboratory, PO Box 550, 65 River Road, Truro, NS B2N 5E3,
902-893-6538

Ontario

Poultry Specialist, Department of Animal and Poultry Science, University of
Guelph, Guelph, ON N1G 2W1, 519-824-4120
Veterinary Laboratory, Guelph Agriculture Centre, Ontario Veterinary College,
Guelph, ON N1G 6N1 519-823-8800

Prince Edward Island

Atlantic Veterinary College, University of Prince Edward Island, 550 University
Avenue, Charlottetown, PEI C1A 4P3, 902-566-0967

Quebec

Department of Veterinary Pathology, School of Veterinary Medicine, CP 5000,
St. Hyacinthe, QC J2S 7C6, 514-773-8521
Poultry Specialist, Department of Animal Science, Macdonald College, McGill
University, Ste. Anne de Bellevue, QC H9X 1C0, 514-398-7707

Saskatchewan

Department of Veterinary Pathology, Western College of Veterinary Medicine,
University of Saskatchewan, Saskatoon, SK S7N 0W0, 306-966-7299
Poultry Specialist, Department of Animal and Poultry Science, University of
Saskatchewan, Saskatoon, SK S7N 0W0, 306-966-7299

ORGANIZATIONS

In addition to the national organizations listed below, many breeds are sponsored by national or regional organizations, and many states and local areas have clubs for exhibitors.

American Egg Board, 1460 Renaissance Drive, Park Ridge, Illinois 60068, 847-296-7043. Promotes egg consumption.

American Livestock Breeds Conservancy, PO Box 477, Pittsboro, North Carolina 27312, 919-542-5704. Encourages the preservation of rare dual-purpose breeds in the U.S.A.

American Poultry Association, c/o Lorna Rhodes, 133 Millville Street, Mendon, Massachusetts 01756, 508-473-8769. Promotes the raising and breeding of exhibition fowl of all sorts.

International Chicken Flying Association, Bob Evans Farm, PO Box 330, Rio Grande, Ohio 45674, 614-245-5305. Sponsors chicken-flying and rooster-crowing contests.

Rare Breeds Canada, Trent University Environmental and Resource Studies Program, Petersbourgh, CN K9J7B8 Canada, 705-748-1634. Encourages the preservation of rare dual-purpose breeds in Canada.

Society for the Preservation of Poultry Antiquities, c/o Marion Nash, PO Box 102, 210 South 10th Street, Murphysboro, Illinois 62966-0210, 618-684-3710. Encourages the preservation of rare exhibition breeds.

RECOMMENDED READING

Some of the publications listed below are regularly updated, so be sure to look for the latest editions of those with that designation.

American Poultry History 1823-1973 (1974), John L. Skinner, editor, American Poultry Historical Society, Inc., Poultry Science Department, University of Wisconsin, Madison, WI 53706, 608-262-1243. Chronicles the development of poultry production in the U.S.A. from early exhibitors and backyard enthusiasts to modern commercial farming.

American Standard of Perfection (latest edition), American Poultry Association, c/o Karen Porr, 72 Springer Lane, New Cumberland, PA 17070, 717-774-1926. Pictorial guide for breeders and exhibitors of fancy fowl.

Bantam Standard (latest edition), American Bantam Association, PO Box 127, Augusta, NJ 07822. Pictorial guide for breeders and exhibitors of bantams.

A Chick Hatches (1976), Cole and Wexler, William Morrow (out of print, check your local library or interlibrary loan). Graphic photographs of an embryo at various stages of incubation.

Chicken Diseases (1978), F. P. Jeffrey, American Bantam Association, PO Box 127, Augusta, NJ 07822. Booklet on backyard disease prevention and control.

The Chicken Health Handbook (1994), Gail Damerow, Storey Communications, Schoolhouse Road, Pownal, VT 05261, 800-441-5700. Comprehensive book on preventing, identifying, and treating diseases in backyard flocks.

Chicken Tractor (1994), Andy Lee, Good Earth Publications, PO Box 898, Shelburne, VT 05482, 802-425-3201. Ideas for combining chickens with gardening for soil improvement and/or small-farm income.

Cooping In (1995), Pat Rubin, Talisman Press, PO Box 5485, Auburn, CA 95604. 24-page guide to grooming and showing.

Eggcyclopedia (latest edition), American Egg Board, 1460 Renaissance Drive, Park Ridge, IL 60068, 708-296-7043. Booklet filled with fascinating and fun facts about eggs.

Egg-Grading Manual (latest edition), Agriculture Handbook #75, Poultry Division, AMS, USDA, P.O. Box 96456 Washington, DC 20090-6456, 202-720-4476. Illustrated guide and color wall chart; accompanying film available through your county Extension agent or the film library at your state land-grant university.

Fancy Fowl, c/o Elaine Nelisse, 11067 Madison Road, Montville, OH 44064, 216-968-3381 (or Crondall Cottage, Highclere, Newbury, Berks RG15 9PH, Great Britain). Bimonthly color magazine on exhibition and rare breeds, published in England.

The Feather Fancier, Jase Greiner, Editor, Rural Route 5, Lake Road, Forest, ON, CANADA N0N 1J0, 519-899-2364. Monthly newspaper covering Canadian exhibition poultry.

Fences for Pasture & Garden (1992), Gail Damerow, Storey Communications, Schoolhouse Road, Pownal, VT 05261, 800-441-5700. Thorough guide to designing and constructing fences of all kinds.

Free-Range Poultry (1990), Katie Thear, Farming Press, Diamond Farm Enterprises, PO Box 537, Alexandria Bay, NY 13607, 800-481-1353 (or Farming Press Books, Wharfedale Road, Ipswich 1P1 4LG, Great Britain). British manual on small-scale commercial free-range meat and egg production.

Greener Pastures on Your Side of the Fence (1991), Bill Murphy, Arriba Publishing, 212 Middle Road, Colchester, VT 05446, 802-878-2347. Comprehensive book about pasture management for all types of livestock including poultry.

Grit and Steel, Drawer 280, Gaffney, SC 29342. Monthly magazine for devotees of sport Game fowl.

Information Bank, American Bantam Association, PO Box 127, Augusta, NJ 07822. Series of pamphlets on showing and breeding bantams; send self-addressed stamped long envelope for list.

Information Packages, Appropriate Technology Transfer for Rural Areas, PO Box 3657, Fayetteville, AR 72702, 501-442-9824, 800-346-9140. Descriptions of latest technology on natural (organic) poultry and egg production, sustainable (free-range) poultry production, and marketing.

A Manual of Poultry Diseases (latest edition), Department of Agricultural Communications, Reed McDonald Building, Room 101, Texas A&M University, College Station, TX 77843. Handbook describing common diseases and their control.

National Poultry Improvement Plan Directory of Participants Handling Egg-Type and Meat-Type Chickens and Turkeys (latest edition), USDA, APHIS Veterinary Services, 4700 River Road, Riverdale, MD 20737-1231, 202-720-4476. List of certified meat or egg flock owners, dealers, and hatcheries by state.

National Poultry Improvement Plan Directory of Participants Handling Waterfowl, Exhibition Poultry, and Game Birds (latest edition), USDA, APHIS-VS, NPIP, 1500 Klondike Road, A102, Conyers, GA 30207, 404-922-3496. List of certified exhibition flock owners, dealers, and hatcheries by state.

Nutrient Requirements of Poultry (latest edition), National Research Council, National Academy Press, 2101 Constitution Avenue, NW, Washington, DC 20418, 202-334-3313. Technical discussions and feed charts for formulating rations.

Old Poultry Breeds (1989), Fred Hams, Shire Publications Ltd., available from American Livestock Breeds Conservancy, PO Box 477, Pittsboro, NC 27312, 919-542-5704. British booklet outlining the development of various breeds.

The Other Half of the Egg (1967), Helen McCully et al., M. Barrows & Company, Inc. (out of print, check your local library or interlibrary loan). Unique cookbook offering ideas for using leftover egg yolks or whites.

Pastured Poultry Profit$ (1993), Joel Salatin, Polyface, Inc., Route 1 Box 281, Swoope, VA 24479, 703-885-3590. A manual on commercial-scale free-range meat and egg production.

Poultry Census and Sourcebook (latest edition), American Livestock Breeds Conservancy, PO Box 477, Pittsboro, NC 27312, 919-542-5704. Descriptions of and sources for rare dual-purpose varieties.

Poultry Health Handbook (1988), L. Dwight Schwartz, Publications Distribution Center, College of Agricultural Sciences, The Pennsylvania State University, 112 Agricultural Administration Building, University Park, PA 16802. Guide to identifying and preventing diseases of chickens and other fowl.

Poultry Press, William F. Wulff, Editor, PO Box 542, Connersville, IN 47331, 317-827-0932. Monthly newspaper reporting on poultry shows in the U.S.A.

Raising Rare Breeds (1994), Jy Chiperzak, Joywind Farm Rare Breeds Conservancy, Inc., General Delivery, Campbellford, Ontario, K0L 1L0 Canada, 705-653-0231. Spiral bound manual of breeding strategies for small flocks (and other livestock).

Regulations Governing the Voluntary Grading of Poultry Products, and U.S. Classes, Standards, and Grading (7 CFR Part 70), United States Department of Agriculture, Washington, DC. Periodically updated regulations pertaining to standards for marketing dressed meat birds (may be available through your state poultry specialist or agricultural library).

Your Chickens: A Kid's Guide to Raising and Showing (1993), Gail Damerow, Storey Communications, Schoolhouse Road, Pownal, VT 05261, 800-441-5700. A guide to the basics of chicken raising and showing for children aged 9 and up.

Videos

Feeding the Flock, Agricultural Information Department, Drawer 3AI, New Mexico State University, Las Cruces, NM 88003.

Flock Management, Home Poultry Flock Management, Quality Eggs from Small Flocks, Managing Your Tabletop Incubator, How to Slaughter and Process Poultry, Publications Distribution Center, College of Agricultural Sciences, The Pennsylvania State University, 112 Agricultural Administration Building, University Park, PA 16802.

The Growing Embryo, Film Library, New York State College of Agriculture, Cornell University, Ithaca, NY 14850.

Introduction of Flock Health, Central Mailing Services, Oklahoma State University, Stillwater, Oklahoma 74078.

Poultry at Home, Diamond Farm Enterprises, PO Box 537, Alexandria Bay, NY 13607 *or* Farming Press Videos, Wharfedale Road, Ipswich 1P1 4LG England.

Suppliers

The following sources of stock, equipment, and veterinary supplies are listed for your convenience. No endorsement is expressed or implied.

Collapsible Wire Products
5120 N 126 Street
Butler, WI 53007
414-781-6125
Wire exhibition cages.

Cutler's Supply
3805 Washington Road
Carsonville, MI 48419
810-657-9450
General supplies.

First State Veterinary Supply
PO Box 190
Parsonburg, DE 21849
800-950-8387
Medications.

G.Q.F. Manufacturing Company
2343 Louisville Road, PO Box 1552
Savannah, GA 31498
912-236-0651
Wire pens, incubators, brooders, supplies.

The Humidaire Incubator Company
217 West Wayne Street, PO Box 9
New Madison, OH 45346
513-996-3001
Redwood cabinet incubators.

Inman Hatcheries
PO Box 616
Aberdeen, SD 57402
800-843-1962
Chicks of various breeds.

Jeffers Vet Supply
PO Box 100
Dothan, AL 36302
800-533-3377
Medications and general supplies.

Keipper Cooping Company
Dale N. Reimer
W224 S8475 Industrial Drive
Big Bend, WI 53103
414-662-2290
Wire exhibition and training coops.

Lyon Electric Company
2765 Main Street
Chula Vista, CA 91911
619-585-9900
Incubators, brooders, parts, and accessories.

Max-Flex
U.S. Route 219
Lindside, WV 24951
304-753-4387, 800-356-5458
Electroplastic fencing.

Murray McMurray Hatchery
Webster City, IA 50595
515-832-3280, 800-456-3280
Chicks, general supplies, and books.

Nasco Farm & Ranch
901 Janesville Avenue
Fort Atkinson, WI 53538-0901
414-563-2446, 800-558-9595
Full line of general supplies.

Omaha Vaccine Company
PO Box 7228, 3030 "L" Street
Omaha, NE 68107
800-367-4444
Medications and supplies.

Patterson Poultry Supplies
210 Meadowbrook Lane
Martinsville, VA 24112
540-638-2297
Full line of supplies and books.

Poultry Ephemera
Sheila Holligon
Brookside Farm
Fryup, Lealholm, Whitby, Yorkshire,
England YO21 2AP
British books.

Premier Fence Supply
2031 300th Street
Washington, IA 52353
319-653-7622, 800-282-6631
Electroplastic fencing.

Safeguard Products, Inc.
PO Box 8
New Holland, PA 17557
800-433-1819
*Wire cages, components, tools,
accessories.*

Sand Hill Preservation Center
1878 230th Street
Calamus, IA 52729
319-246-2299
Chicks in rare and heirloom breeds.

Smith Poultry & Game Bird Supply
14000 West 215th Street
Bucyrus, KS 66013
913-879-2587
Books and general supplies.

Strecker Supply Co.
PO Box 190 Dept. S
Parrisonburg, MD 19975
800-765-0065
Medications.

**Stromberg's Chicks and Game Birds
Unlimited**
PO Box 400
Pine River, MN 56474
218-587-2222
Chicks, full line of supplies and books.

Tomahawk Live Trap Co.
PO Box 323
Tomahawk, WI 54487
715-453-3550
Humane predator traps.

Val-A Chicago
700 W Root Street
Chicago, IL 60609
773-927-9442, 800-552-8252
General supplies.

Waterford Corporation
404 North Link Lane
Fort Collins, CO 80524
970-482-0911, 800-525-4952
Electroplastic fencing.

INDEX

Illustrations are indicated by page numbers in *italics;* charts and tables are indicated by pages numbers in **bold.**